MW00849390

STARBOUND

This book is for anyone enthralled by the romantic dream of a voyage "to the stars." From our current viewpoint in the twenty-first century, crewed interstellar travel will be an exceptionally difficult undertaking. It will require building a spacecraft on a scale never before attempted, at vast cost, relying on unproven technologies. Yet somehow, through works of science fiction, TV, and movies, the idea of human interstellar travel being easy or even inevitable has entered our popular consciousness. In this book, Ed Regis critically examines whether humankind is bound for distant stars, or if instead we are bound to our own star, for the indefinite future. How do we overcome the main challenge that even the nearest stars are unimaginably far away? He explores the proposed technologies and the many practical aspects of undertaking an interstellar journey, finishing with his reflections on whether such a journey should be planned for.

Ed Regis holds a PhD in philosophy from New York University and is the author of 10 books about science and technology. He has also written for *Scientific American*, *Wired*, *Nature*, *Harper's*, *Omni*, *Discover*, and *Air & Space Smithsonian*. He lives with his wife, Pam, in the Maryland mountains.

Starbound

Interstellar Travel and the Limits of the Possible

ED REGIS

Shaftesbury Road, Cambridge CB2 8EA, United Kingdom

One Liberty Plaza, 20th Floor, New York, NY 10006, USA

477 Williamstown Road, Port Melbourne, VIC 3207, Australia

314–321, 3rd Floor, Plot 3, Splendor Forum, Jasola District Centre, New Delhi – 110025, India

103 Penang Road, #05–06/07, Visioncrest Commercial, Singapore 238467

Cambridge University Press is part of Cambridge University Press & Assessment, a department of the University of Cambridge.

We share the University's mission to contribute to society through the pursuit of education, learning and research at the highest international levels of excellence.

www.cambridge.org
Information on this title: www.cambridge.org/9781009457590

DOI: 10.1017/9781009457552

© Ed Regis 2025

This publication is in copyright. Subject to statutory exception and to the provisions of relevant collective licensing agreements, no reproduction of any part may take place without the written permission of Cambridge University Press & Assessment.

When citing this work, please include a reference to the DOI 10.1017/ 9781009457552

First published 2025

Printed in the United Kingdom by CPI Group Ltd, Croydon CR0 4YY

A catalogue record for this publication is available from the British Library

A Cataloging-in-Publication data record for this book is available from the Library of Congress

ISBN 978-1-009-45759-0 Hardback

Cambridge University Press & Assessment has no responsibility for the persistence or accuracy of URLs for external or third-party internet websites referred to in this publication and does not guarantee that any content on such websites is, or will remain, accurate or appropriate.

For Pamela

CONTENTS

PREFACE

> When a distinguished but elderly scientist states that something is possible, he is almost certainly right. When he states that something is impossible, he is very probably wrong.
>
> Arthur C. Clarke, *Profiles of the Future* (1964)

It is no part of the purpose of this book to argue that human interstellar travel is impossible (although it might in fact be). The title, *Starbound*, is deliberately ambiguous as between (a) *we are bound for the stars* in the sense that humans are one day going to exit our solar system and head for a planet orbiting a star in another solar system light years away; and (b) *we are bound to our own star*, the Sun, for the indefinite future. It is the object of this book to explore the issues surrounding interstellar travel in such a way that will allow the reader to come to a decision as to which of these two alternatives is the more likely.

It is also no part of the purpose of this book to argue that interstellar travel is easy, or that it is going to happen anytime soon. The path to the stars is littered with obstacles at each and every turn, and it is by no means obvious that they can all be overcome, avoided, or otherwise circumvented. Indeed, there are plenty of reasons for thinking that even if it's possible in principle, traveling to the stars is nevertheless extremely difficult and is therefore improbable in a practical sense. But this doesn't mean that a human flight to the stars will never take place. Progress in science and technology has often converted what was once regarded as improbable, or even as impossible, into practical, everyday, working realities. Human flight, transatlantic flight, human beings on the Moon, these and all the other modern miracles were each considered as impossible – until they happened.

Nevertheless, there are good reasons for thinking that star travel is improbable in practice. One of them is that no flight to even the nearest stars can be made, with current propulsion technology, within a human lifetime. This is because current spaceflight technology is limited to chemical propulsion systems, and such systems are not powerful enough to accelerate an interstellar vehicle to a speed that would take us to the stars within a normal human lifespan. The core problem is that the stars are so far away.

Even the "nearest" star, Proxima Centauri, is unimaginably far away: it is 4.246 light years from Earth, which translates to 25,300,000,000,000 miles, or 25.3 trillion miles. If we were to travel as fast as the Voyager I spacecraft, which left the solar system in September 2013, and which was then moving at the rate of 38,698 miles per hour, it would take 73,000 years to get from here to there.

There are two possible solutions to this, however. The first and most obvious is simply to go faster. The fastest spacecraft in history is the NASA Parker Solar Probe, which reached the speed of 364,660 miles per hour on November 20, 2021. At that speed, it would still take a spacecraft almost a thousand years to get to Proxima Centauri. But the Parker Solar Probe is about the size of a small car, whereas the starships envisioned by interstellar travel advocates are immense.

Attempts to go even faster, say at 10 percent the speed of light, run into propulsion and propellant difficulties. Ten percent the speed of light is an almost incomprehensible velocity: it's equivalent to 18,628 miles per *second*, or 67,616,629 miles per hour. Traveling at 67 million miles per hour would allow us to get to Proxima Centauri in a little more than 40 years, which is well within a human lifetime, but it is still a long time for a human crew to be kept inside a closed habitat hurtling through space. The problem here, however, is that current rocket technology is not capable of propelling a starship to such speeds. Physicist Richard Obousy, past president of the advocacy firm Icarus Interstellar, has calculated that getting a starship to achieve

10 percent the speed of light would require more chemical propellant than exists on Earth today.

And so interstellar travel advocates are forced to enter the realm of "alternative" or "exotic" propulsion systems and propellants. These are systems that don't exist today, and are possible only in theory – if that. They are systems and devices that exist only notionally or hypothetically, as *ideas*. Antimatter propulsion, for example, is a scheme by which normal matter and antimatter are brought together and annihilate each other, releasing enormous amounts of energy, which is then converted into thrust.

That plan, however, faces a number of obstacles. One, since antimatter destroys conventional matter, it would be impossible to store it in normal physical containers. The solution, though, is to store it inside a strong magnetic field, which can be, and in fact has been done. It would also be impossible to direct antimatter thrust by means of conventional, physical exhaust nozzles such as are used in chemical rocketry, but it is possible to vector the exhaust in a given direction by means of the same magnetic field. And finally, the amount of antimatter that has been created so far in particle accelerators is infinitesimal, and has been produced at great expense. It may be possible to scale up antimatter production, and also to reduce the cost of doing so, although it is an open question whether antimatter could be mass-produced in quantities sufficient to power a spacecraft across interstellar distances.

Another advanced propulsion concept is to use controlled nuclear fusion to drive the spacecraft. Controlled nuclear fusion is currently under active development, and has been achieved more than once in the form of short bursts of energy. But the energy pulses so far produced are so very short – in the trillionths or billionths of a second – as to make it questionable whether controlled fusion could be sustained for the time period long enough to propel a spacecraft to the stars.

And so we turn then to the second possible solution to the distance problem, which is to build a spacecraft that could accommodate several generations of humans in succession over the span of many decades, or even centuries, of travel time. Such spacecraft are known as multigenerational vehicles, world ships, or interstellar arks. But designing and building a multigenerational spacecraft would be an engineering feat on a scale of size and complexity never before attempted, and not known to be possible. Even if we assume that building such a craft is possible in principle, the engineering challenges posed by the craft's sheer bulk and complication might be daunting. But engineering challenges can be overcome, even if it is extremely difficult, expensive, and time-consuming to do so.

The idea of a multigenerational spacecraft also raises a moral problem, at least for the generations that will have been born upon the ship. The problem is that none of those to be born on the spacecraft will have given their consent to be confined inside it for their entire lifetimes. The only generation that will have given their consent is the crew that initially boards the ship, who will do so voluntarily; none of the successive follow-up generations will have given their free and informed consent to be there. It is true that no one born on Earth gave their consent to be born there, either, and we do not see that as morally problematic. But are the two situations similar enough as to make them morally equivalent?

Suppose, however, that the moral problem has been resolved or otherwise sidestepped, that a sufficiently powerful propulsion system has been developed, and that a multigenerational starship was built, staffed, launched, and reached a relativistic velocity on the order of 10 percent of light speed. Then a new problem arises, which is that the possibility of the ship's colliding with even a small speck of interstellar matter becomes a deadly threat. Physicist Tom W. Gingell of Science Applications International Corporation, who did a study of the subject, found that at such speeds an impact with even a random small particle would have the effect upon the spacecraft of an H-bomb explosion. But in his study, "Starship Collision Warning Using Quantum Radar" (2012), Gingell also suggested a possible solution. And this was to use quantum radar (which makes use of entangled microwaves) to detect the incoming object, and then utilize an automated defense system to deflect or destroy it. Other researchers have proposed flying a thick metal shield miles ahead of the spacecraft in order to accomplish the same deflection or destruction of the threatening object.

But even if we could make a journey to the stars safely, the question arises whether the trip would be worth the cost. For the reason most commonly given for traveling to the stars, which is to preserve the human species if and when conditions on Earth make the planet unlivable, can be fulfilled more easily, quickly, safely, and cheaply simply by going elsewhere within the solar system. Possible destinations include the Moon and Mars, and satellites such as Europa or Enceladus. Another option would be to create O'Neill-type space colonies in Earth orbit, or among the asteroids, or anywhere else within our own system of planets.

Is there anything uniquely specific or attractive about proximity to a distant star system that would make such a location more suitable for the preservation of the human species than a location elsewhere in our own solar system? Yes, in light of the fact that sooner or later our Sun will become a red giant and will engulf the planets of the inner solar system. Still, this will not happen for billions of years, and so it gives us no reason to leave for the stars now, or at any time in the near future.

As this brief survey suggests, there are many barriers to overcome before human interstellar travel becomes a realistic, practical, and desirable proposition. It is the purpose of this book to explore the historical and intellectual development of interstellar spaceflight theory, and to assess the plausibility of the major systems, spacecraft, and related technologies that have been proposed for getting us to the stars.

Even in the face of all the difficulties mentioned above (and perhaps still others to come), we have seen no reason to think that interstellar travel is impossible in principle, or that it will never happen – ever. A journey to the stars does not appear to violate any known law of nature. It is therefore entirely possible *in theory*. The question is whether it is possible *in practice*, in a reasonable, doable, feasible, real-world, and practical sense. That is the question to be addressed at length in this book.

1 ORIGINS OF THE DREAM

Traveling to the stars is one of the greatest and grandest dreams of humanity. The prospect of leaving the Earth behind, blasting through light years of empty space, reaching another solar system, and then alighting upon a completely new and alien world, is one of the biggest, boldest, and bravest schemes ever conceived. The idea pervades popular culture, having been depicted in a long line of science fiction short stories and books, in television series such as *Star Trek* and its sequels and imitators, and in movies like *Star Wars* and *Interstellar*, plus their own train of sequels, copies, and spinoffs. As a result, it is all too easy for us to imagine that one day – maybe even one day soon – some select portion of the human race will actually be *leaving for the stars*.

And what a romantic dream it is! The whole enthralling vision is magnetically attractive and intoxicating. As Daniel Deudney said in his book *Dark Skies* (a critique of space expansionism): "It feeds off experiences of the technological and cosmic sublimes, of awesome roaring rockets and astounding gizmos, and of weirdly alien planets, incomprehensibly titanic cataclysms, and mesmerizing spacescapes. Everything about it is oversized: speeds, distances, magnitudes – and payoffs."

For all of their appeal, however, dreams have their limitations. They are expressions of emotional states, desires, and yearnings as opposed to being cognitive structures embodying practical human interests, needs, or necessities. To say this is not to downplay or dismiss the value of dreams. Dreams have a legitimate, perhaps even a necessary, place in the overall scope of human aspiration, creativity, and achievement. Dreams motivate people to attempt, and to accomplish, great feats of construction, such as an architect building the world's tallest skyscraper or an engineer designing a rocket that will take us to the planet Mars. History's biggest megaprojects – the Suez Canal, the Channel Tunnel, the Apollo Project – had been dreams in the minds of their respective planners and engineers long before they were ever begun.

People have even discovered literal truths in dreams. In the mid-1800s the German chemist Friedrich August Kekulé happened upon two of his most important discoveries while dreaming, and later recorded a detailed account of each dream. The first discovery was his idea of the manner in which carbon atoms combined with others to form carbon compounds. The solution occurred to him while riding an open-top bus in London:

> One fine summer evening I was returning by the last omnibus, "outside," as usual through the deserted streets of the metropolis, which are at other times so full of life. I fell into a reverie, and lo, the atoms were gambolling before my eyes!. . . . The cry of the conductor "Clapham Road" awakened me from my dreaming; but I spent part of the night in putting on paper at least sketches of these dream forms.

Kekulé's better-known discovery was that of the hexagonal structure of the benzene molecule, which likewise came to him in a dream:

> I was sitting, writing at my text-book; but the work did not progress; my thoughts were elsewhere. I turned my chair to the fire and dozed. Again the atoms were gambolling before my eyes. This time the smaller groups kept modestly in the background. My mental eye, rendered more acute by repeated visions of the kind, could now distinguish larger structures, of manifold conformation: long rows, sometimes more closely fitted together; all twining and twisting in snake-like motion. But look! What was that? One of the snakes had seized hold of its own tail, and the form whirled mockingly before my eyes. As if by a flash of lightning, I awoke; and this time also I spent the rest of the night in working out the consequences of the hypothesis.

These experiences gave Kekulé a new respect for dreams. "Let us learn to dream," he said, "then perhaps we shall find the truth."

# # #

A dream so vast, complex, and multifaceted as that of traveling to the stars has, like a river flowing from many tributaries, more than one source or origin. In addition, the sources may be of different types, originating as myths, as desires, in abstract conceptual thought, from the construction of proofs of concept, and in the results of experimental tests. As a matter of simple logic, the first and most basic prerequisite of interstellar travel was the prior notion of human flight through the air. The dream of leaving the surface of the Earth and traveling through the atmosphere is quite ancient. It originated in myth, and then had a somewhat tortuous developmental history that was littered with a number of dead ends as well as an even greater number of dead people.

The idea is reflected in many legends going back to antiquity, including the myth of Daedalus and Icarus, which is the best known among them. But there are also other myths, from earlier times and different cultures. In his book *Taking Flight*, the aviation historian Richard Hallion said:

> From the dawn of time people around the globe have expressed the dream of flight, emphasizing the incredible and depicting aerial powers as an element of religion, mythology, or war. The Egyptians worshipped Horus, the sky-god falcon ... The peoples of Asia Minor venerated various flying deities, such as the winged Hurrian goddess Shaushka (Ishtar) and the sacred double-headed eagle of Hittite tradition.

And the Christians, Jews, and Muslims had their divine winged angels.

But even before people had any thought of taking to the skies, they had invented all sorts of gadgets and contrivances that flew, and their designs reflected at least a primitive, intuitive understanding of what later became known as the principles of aerodynamics. Even a child playing games had a natural grasp of how to throw stones or branches so as to make them reach their targets. A more adult activity was to propel straight, elongated objects swiftly and accurately, whether in outdoor games or as weapons that might mean the difference between eating or starving – or even life or death, depending on the circumstances. Thus there are ancient spears and arrows that are efficiently streamlined, that do not tumble in flight, and that follow the same flight path when launched in the same manner, allowing for the development of marksmanship.

The mythical flight of Daedalus and Icarus was patterned after the flight of birds, for both fliers had feathers attached to their arms. Their flight was in fact a partial success since although Icarus lost his feathers by flying "too close to the sun," and plunged into the sea and drowned, his father Daedalus landed safely on the shore.

The idea of using birds as templates for human flight was attractive inasmuch as birds were an obvious proof of concept that heavier than air flight was possible, and for that reason it was embraced by many who hoped to get people aloft. So great a person as Leonardo da Vinci had closely observed the flight of birds, and wrote in his notebook: "A bird is an instrument working according to mathematical law, which instrument it is within the capacity of man to reproduce with all its movements." He also thought the same about bats: "Dissect the bat, study it carefully, and on this model construct the machine."

This was the concept of an ornithopter, a mechanical device propelled by a person equipped with wings that could be flapped by muscle power. Da Vinci sketched a number of designs for such contrivances, all of which were essentially unworkable in practice. Birds are relatively small and light, and the members of many species have hollow bones. The largest and heaviest birds, such as ostriches and emus, are flightless. The best and most agile fliers – hummingbirds – are so small and light as to weigh almost nothing at all: the bee hummingbird, of Cuba, weighs less than half an ounce.

Human beings are far too heavy to be supported in the air by means of the lift provided by flapping wings, as several men who built ornithopters and tried to fly in them demonstrated. Between 850 and 1500, various "tower jumpers" built wings out of wood, feathers, or cloth, and then jumped off roofs, trees, and other heights, desperately flapping away, often killing themselves in the process. The bird template therefore constituted a dead end in the attempt to make human flight possible, and retarded rather than accelerated the development of flying machines.

But amid the history of failed bird-mimicking flight concepts and actual flight attempts, there appeared in the first decade of the eleventh century an instance of true winged flight that must be regarded as at least a partial success. This was the case of a young Benedictine monk named Eilmer, of Malmesbury Abbey in England. The abbey, portions of which still exist, had walls that were about 150 feet high (about as tall as a 15-story building), and is located in western Wiltshire, an area characterized by frequent gusty winds.

Brother Eilmer had been inspired by the Daedalus legend, and also perhaps by the profusion of crows that nested upon the building's heights, and so constructed for himself a crude set of wings that could attach to his hands and feet. The wings are thought to have been made of wood covered by a light cloth. The outstretched wings did not flap, but in the air they would be pushed upward at their tips, giving them a dihedral (V-shaped) angle with respect to each other, a geometrical arrangement that acted to keep the wings laterally stable in flight.

And so one fine day, probably when the wind was gusting favorably, in an act of singular courage (and considerable recklessness), Brother Eilmer climbed to the top of the abbey wall and stood on the ledge for a moment, balancing himself. He looked down upon the tops of trees below him, as well as the river in the distance, and perhaps even saw a flock of birds. And then, no doubt murmuring a few words of prayer, launched himself off into empty space.

He did not fall downward like a stone, but, according to a later historian of British monks, William of Malmesbury, glided a fair distance away. As William described the incident:

> [Eilmer] had by some contrivance fastened wings to his hands and feet, in order that, looking upon the fable as true, he might fly like Daedalus, and collecting the air, on a summit of the tower, had flown for more than a furlong; but agitated by the violence of the wind and a current of air, as well as by the consciousness of his rash attempt, he fell and broke his legs, and was lame ever after. He used to relate as the cause of his failure that he had forgotten to provide himself with a tail.

A furlong was about 600 feet, approximately the length of two football fields, and to travel through the air that far horizontally was a genuine example of gliding flight even if the landing came with a bang. His glide path was relatively steep, and the whole flight is thought to have taken between 12 and 15 seconds. But despite its painful ending, Brother Eilmer's floating descent to the ground was an early, real-life example of hang gliding.

# # #

The first true, safe, and successful free ascent of a human being from the ground and into the air, and back, was not achieved by an ornithopter, hang glider, or anything else having to do with wings. It was made, instead, by . . . a hot-air balloon.

One might imagine that the first humans to successfully achieve the age-old dream of flight would be household names, familiar to all the world, but in fact they are complete unknowns to most people today. But the balloon by which they flew was an invention of the comparatively well-known Montgolfier brothers, Étienne and Joseph, papermakers, of Annonay, France. The idea for the craft had originated with Joseph, who in November 1782 sat gazing into a fireplace. He saw little ash particles being carried up the flue and came to the sudden realization that this rising current of heat might be captured in some manner and be put to practical use.

He soon constructed a small test balloon out of taffeta, filled it with smoke from a fire, and lo and behold, it rose into the air. The two brothers then constructed a bigger test model, the *Martial*, which in September 1783 made an ascent witnessed by Louis XVI, Marie Antoinette, and an estimated 100,000 Parisians. In what would become the grand tradition of putting animals inside spacecraft before using human subjects, in the balloon's basket were a sheep, a duck, and a rooster.

After that success, the Montgolfier brothers created their masterpiece vehicle of flight, the *Globe Aërostatique*. Looking much like a flying Fabergé egg, it was said to be one of the most richly decorated aerial vehicles ever to exist. Its outer surface was covered with the 12 signs of the Zodiac, the King's initials, multiple radiating sunbursts, a repeating pattern of fleurs-de-lys, plus representations of draperies, bunting, and various other oddments and adornments.

The test pilot for the balloon's maiden flight would be one Jean François Pilâtre de Rozier, a 26-year-old *citoyen*-scientist and adventurer who was always ready for anything. On Friday, November 21, 1783, Rozier and his co-pilot, François Laurent, the Marquis d'Arlandes, boarded the craft in the Bois de Boulogne. With heat provided by a brazier burning straw and wool, at 1:54 in the afternoon the blue and gold balloon rose from the ground "in a most majestic fashion," according to an observer, gained height to the degree that it could be seen by all of Paris, and flew off toward the Seine. Twenty to 25 minutes later, the aeronauts landed safely in a field on the outskirts of the city. So ended the first human flight in the history of the world.

The balloon in its various incarnations proceeded to have a checkered career, carrying passengers, and often killing them, until it all came to a tragic conclusion in the *Hindenburg* disaster, in which the enormous Zeppelin crashed and burned at Lakehurst, New Jersey, on May 6, 1937. Thirty-five people out of a total of 97 souls on board perished in the flames or died in hospital later.

Lighter-than-air flight lives on today in the colorful sport of hot-air ballooning and in a collection of advertising blimps that fly over crowds at large events. As such, the balloon proved to be only a small step up the conceptual ladder that led – at last, and eventually – to the dream of flying to the stars.

# # #

The most significant next step up the ladder was an enormous one, taken by the Wright brothers, who made their famous first flights on December 17, 1903. It was the start of a progression that culminated in the development of the modern, safe, and efficient airliner that we know today. The Wright brothers had proceeded slowly and scientifically, experimenting first with kites and then with gliders. They constructed two wind tunnels in which they tested more than 200 airfoils of different cambers and chords, establishing coefficients of lift for each. Only after years of such research and trial and error did they achieve actual flight.

Still, flying through the atmosphere of Earth, as astonishing as it was at first, constituted only one stage of many leading to the dream of traveling to the stars. A further necessary concept would require a complete change of orientation and frame of reference, for this was the notion of not only leaving the surface of Earth, but of departing from both the planet and its atmosphere, and traveling to another heavenly body, the closest of which was of course the Moon. Surprisingly, this quite radical and literally outlandish notion was nevertheless also quite ancient. When it first appeared, however, the idea was not meant to be taken seriously but was rather intended only as part of an elaborate fiction, or even as a sort of hoax or joke.

The author of this fable was one Lucian (circa 150 CE) of Samosata, a town in what is now Turkey. Lucian was a professional satirist, lampooner, and spinner of tales, most of them wholly unbelievable. He is credited with having produced more than 80 works of satire and fiction, all of which made gleeful fun of everything and everybody. In a work called *Philosophies for Sale* he depicts the Greek god Zeus as establishing a slave market in which he puts several famous philosophers up for auction, including Pythagoras, Diogenes, Heraclitus, and Socrates. In *The Lover of Lies*, he satirizes belief in paranormal phenomena such as miracle cures for various illnesses. He also ridicules superstition, religion, and the Greek gods. Nothing and nobody was sacred to him, and his works were highly prized for their irreverence and absurdity.

Lucian's most famous work, *Vera Historia*, "The True History," begins by telling the reader that everything in the narrative is completely made-up and false, and should not be believed, despite its title conveying exactly the opposite impression. The fictional history was Lucian's attempt to out-do what he saw as some of the more fanciful elements in Homer's *Odyssey*, and in this he succeeded all too well.

"The True History," which describes a voyage to the Moon and back, is regarded by some critics as the first work of science fiction. The journey begins at sea, with Lucian and some fellow travelers sailing out beyond the Pillars of Hercules (the Strait of Gibraltar), where they come upon an island endowed with a river that flows with wine, and where the trees look like women.

The online Project Gutenberg translation by Thomas Francklin describes the course of events as the travelers prepare to leave the island:

> We then took our casks, filled some of them with water, and some with wine from the river, slept one night on shore, and the next morning set sail, the wind being very moderate. About noon, the island being now out of sight, on a sudden a most violent whirlwind arose, and carried the ship above three thousand stadia, lifting it up above the water, from whence it did not let us down again into the seas but kept us suspended in mid air, in this manner we hung for seven days and nights, and on the eighth beheld a large tract of land, like an island, round, shining, and remarkably full of light; we got on shore, and found on examination that it was cultivated and full of inhabitants, though we could not then see any of them. As night came on other islands appeared, some large, others small, and of a fiery colour; there was also

below these another land with seas, woods, mountains, and cities in it, and this we took to be our native country.

They soon encounter another individual, who speaks to them in their native language:

"You are Grecians," said he, "are you not?" We told him we were. "And how," added he, "got ye hither through the air?" We told him everything that had happened to us; and he, in return, related to us his own history, and informed us, that he also was a man, that his name was Endymion, that he had been taken away from our earth in his sleep, and brought to this place where he reigned as sovereign. That spot, he told us, which now looked like a moon to us, was the earth.

An unknowing but prophetic word picture of the famous "Earthrise" photo taken from orbit above the Moon by Apollo 8 astronaut William Anders on December 24, 1968 (Figure 1.1).

A second fictional voyage to the Moon was of greater significance, for two reasons. One, it was written by none other than Johannes Kepler, author of the laws of planetary motion. And two, because Kepler's narrative is much more realistic than Lucian's, to the point that some of it rings true, almost as if it were "hard" science fiction or even plain fact. Fittingly enough in this context, the author called his story *Somnium*, "The Dream." A new translation from the Latin of a relevant section by Tom Metcalf of The Somnium Project, conveys a few of the core events during this second fictitious trip to the Moon:

The whole journey, although far, is completed in a time of four hours at the most. Never are we more busy than the time determined for our departure. ... This occasion proves so narrow [just like a space shot!] that we take few of the human race along. ... First of all he experiences a strong pressure, not unlike an explosion of gunpowder, as he is hurled above the mountains and the seas. For this reason, drugs and opium are consumed at the start, so that he falls asleep, and each of his limbs disentangled, so that his body is not torn from his legs [as if by g-forces], nor his head driven from his body, but so the shock will be distributed across all his limbs. Next he experiences new difficulties: it is intensely cold ... When they awake, humans usually complain of indescribable exhaustion in all their limbs, from which, much later, they recover enough to walk.

Strong pressure. G-forces. Intensely cold. Exhaustion on landing and trouble walking – similar enough to what astronauts on long-duration missions on the International Space Station experience when they arrive back on Earth. In his book *Cosmos*, Carl Sagan writes that Kepler got some other things

Figure 1.1 Earthrise. (NASA)

right as well: "Because of the length of the lunar day and night Kepler described 'the great intemperateness of climate and the most violent alteration of extreme heat and cold on the Moon,' which is entirely correct."

The first Project Apollo Moon landing on July 20, 1969 was the real-world embodiment of centuries of fables, hopes, and dreams.

# # #

Two additional conceptual leaps of the imagination were necessary before the dream of interstellar travel could emerge as a realistic possibility. The first was the idea of leaving the Earth–Moon system altogether, and traveling to another planet within the solar system. Lucian of Samosata had mentioned traveling to Venus in his "True Stories," but only as yet another instance of his tall tales. Several other authors also wrote stories and books about traveling to Mars in the late 1800s. But the first to explore this idea at length and in detail, and in such a way as to reach a wide audience, was Edgar Rice Burroughs in his 11 novels about Mars, starting with his first, *The Princess of Mars*, of 1912.

Burroughs's fiction had been inspired in part by the prior work of two astronomers, Giovanni Schiaparelli and Percival Lowell. In 1877 Earth and Mars reached points in their orbits when they were closest together, and therefore made Mars favorable for observation from Earth. During this time Schiaparelli made an exhaustive telescopic study of the Martian surface and decided that he saw there a series of *canali*, meaning channels. The word was mistranslated into English, however, as "canals," implying that they had been deliberately made by intelligent living beings.

In 1894, during another close approach of the two planets, the Harvard educated astronomer Percival Lowell made an even more thorough study of Mars from his own private observatory in Flagstaff, Arizona. He took thousands of photographs of the Martian surface and claimed to have seen complex patterns of canals, which intersected at what he called "oases." He later wrote several works about the planet, including an essay and a book, each entitled "Mars as the Abode of Life." Both claimed that the canals had been built by intelligent beings for purposes of irrigating the planet's otherwise dry land areas. He clarified matters somewhat by saying that "it is evident that what we see, and call by ellipsis the canal, is not really the canal at all, but the strip of fertilized land bordering it – the thread of water in the midst of it, the canal itself, being far too small to be perceptible."

Having promoted visions of an alien, dying, intelligent civilization, Lowell became, according to science writer (and science fiction author) Isaac Asimov, "the patron saint of the intelligent-life-on-Mars cult." Of course no canals existed on Mars, then or now, and Lowell's visions are possibly just wishful projections of his hopes and dreams onto an entirely canal-absent surface. The mystique of Mars nevertheless kept its hold on the general public.

Edgar Rice Burroughs, during his time one of the world's best-selling authors, the creator of *Tarzan of the Apes* among 90 other books, wrote a popular series of novels depicting a complex, intelligent civilization on Mars. The first of them, *A Princess of Mars*, is narrated in the first person by one John Carter of Virginia. It begins strangely enough: "I am a very old man; how old I do not know. Possibly I am a hundred, possibly more; but I cannot tell because I have never aged as other men, nor do I remember any childhood."

At the end of the Civil War he and a fellow officer go prospecting for gold in Arizona. They are quite successful at this, but his colleague ends up being killed by Apache tribesmen. Later, standing alone in an open field in the moonlight, John Carter catches a glimpse of Mars:

> My attention was quickly riveted by a large red star close to the distant horizon. As I gazed upon it I felt a spell of overpowering fascination – it was Mars, the god of war . . . As I gazed at it on that far-gone night it seemed to call across that unthinkable void, to lure me to it, to draw me as the lodestone attracts a particle of iron.

My longing was beyond the power of opposition; I closed my eyes, stretched out my arms toward the god of my vocation and felt myself drawn with the suddenness of thought through the trackless immensity of space. There was an instant of extreme cold and utter darkness.

Next thing he knows, he is on the planet Mars, lying on the ground, naked. "I opened my eyes upon a strange and weird landscape. I knew that I was on Mars; not once did I question either my sanity or wakefulness."

There follows a tale full of what has come to be known as the well-worn chestnuts of science-fictional alien worlds: little green men – "green Martians" – though in this case they were big green men, 15 feet tall. There were also BEMs, bug-eyed monsters: "Their eyes were set at the extreme sides of their heads a trifle above the center and protruded in such a manner that they could be directed either forward or back."

And of course Lowell's canals duly made their appearance: "Twice we crossed the famous Martian waterways, or canals, so-called by our earthly astronomers."

The alluring visions offered by the Burroughs Mars novels – including the magical means of getting there – cast a spell over generations of impressionable young readers, among them scientist Carl Sagan, who confesses in *Cosmos* that "I can remember spending many an hour in my boyhood, arms resolutely out-stretched in an empty field, imploring what I believed to be Mars to transport me there. It never worked."

Sagan would turn out to be one of the world's foremost proponents of colonizing Mars, and of human interstellar migrations. Meanwhile, the first successful landing on Mars occurred on July 20, 1976, when NASA's Viking 1 lander touched down in Chryse Planitia.

# # #

The final leap of the imagination necessary for the dream of traveling to the stars to be entertained as a serious possibility was the biggest conceptual advance of them all: the idea of exiting our native solar system entirely, traveling through not miles but light years of space, landing on a completely new and foreign planet, and establishing a founding civilization there. Whereas the prior feats required only small, incremental progressions from one stage to the next, the act of traveling to the stars was an exponentially greater change of scale, time, and reach.

For this, something more than a dream or a tall tale was needed. What was needed was a genuine spacefaring vehicle, the rocket, as well as an understanding of the principles of space flight, and, lastly, the idea that the true home of humanity was not just the Earth, Mars, or the solar system, but the cosmos.

The first rockets, the essential prerequisites of interstellar travel, depended in turn upon the invention (or accidental discovery) of a propellant, the earliest of which was a derivative of black powder, itself a mixture of sulfur, carbon, and potassium nitrate (saltpeter). The mixture contained an inhibitor so that it would explode slowly and progressively, producing a short, controlled burn rather than a sudden blast. A form of such a chemical propellant appeared in China by the middle of the ninth century. It could be packaged into a length of bamboo that was open at one end; with a stabilizing stick at the bottom and a pointed nose cap at the top, and *voila!* a rocket.

They were at first amusing toys, but soon were put to other uses. Rockets were used in warfare as far back in time as the year 1232, during the Mongol siege of Kaifeng, in China. The technology of black powder rocketry soon made its way to Europe, and it, along with other propellants, have been in use ever afterward.

The idea of using rockets for space flight owes much to the Russian physicist and aerospace engineer Konstantin Tsiolkovsky, who had formulated the basic principles of astronautics before the end of the nineteenth century. According to a short autobiographical sketch, Tsiolkovsky was born in a remote Russian village in 1857. At the age of 10 his hearing had been substantially impaired by a bout of scarlet fever, and he was partially deaf thereafter. "This handicap estranged me from people," he wrote, "and prompted me to read, concentrate, and dream."

One of the things he dreamt about was flight, an interest first stimulated by his mother, who gave him a toy balloon filled with hydrogen. Marvelously, it rose through the air. As he matured, Tsiolkovsky essentially recapitulated the designs of some aeronautical dream-machines that had been proposed by others, starting with the lowly ornithopter and its futile, flapping wings. He built a few examples of these, none of which worked, which brought him to "the final realization that the thing was impractical."

His next step forward, however, was a gigantic one, into what he called "cosmic space." In 1895 he wrote a book, *Dreams about Earth and Skies*, and in 1898 an article, "Reactive Flying Machines." Taken together, these treatises made him one of the fathers of modern space flight. The 1898 essay established the fundamentals of orbital mechanics. It also proposed the then radical idea of using the very same rocket propellants that are still in common use today: liquid oxygen and liquid hydrogen (LOX/LH2). "The two liquid gases are separated by a partition," he wrote. "The place where the gases are mixed and exploded is shown [in an accompanying diagram], as is the flared outlet for the intensely rarefied and cooled vapors."

Later in the same piece Tsiolkovsky advanced the idea of an Earth-orbiting space station: "It is possible to construct a permanent observatory that would travel for an indeterminate length of time around the Earth, like the Moon, beyond the limits of the atmosphere."

Tsiolkovsky later wrote about space suits, satellites, and the colonization of the solar system. At one point he even hinted at the idea of traveling beyond the solar system, saying that "Not only the earth, but the whole universe is the heritage of mankind." And: "Perhaps, a hundred years will pass before my idea will find application and people will travel not only on the surface of our globe but also on the face of the universe." He is popularly known for the aphorism that "Earth is the cradle of humanity, but one cannot remain in the cradle forever."

This, then, was the final culmination of a succession of dreams that had emerged progressively in 11 steps or stages that had begun in antiquity. In logical order, the several steps were from: (1) the birth of ancient Greek and other myths of flight, to (2) proposals for machines that would make flight possible by mimicking the flapping wings of birds, to (3) actual attempts at human flight, to (4) successful human flight through the air by means of balloons, to (5) powered, controlled, sustained human flight through the atmosphere by winged vehicles, to (6) fictional accounts of flying to the Moon, to (7) the invention of rockets leading to an understanding of the principles of space flight, to (8) the Apollo Project Moon landings, to (9) fictional accounts of traveling to Mars, to (10) actual landings on Mars by rockets and robotic rovers, to (11) the idea of leaving Earth and colonizing the universe. Arguably, many of these stages of flight were the embodiments of age-old dreams.

As we have seen, truths can be found in dreams, and Kekulé was a champion of dreaming: "Let us learn to dream, gentlemen, then perhaps we shall find the truth," he said. Still, he ended his praise of dreams with a caveat, on a decidedly cautionary note. "But let us beware of publishing our dreams before they have been put to the proof by the waking understanding."

Can the heroic dream of interstellar flight survive critical scrutiny by the waking understanding, or is it one of those dreams that ends by going up in smoke? That is the question.

2 THE 100 YEAR STARSHIP

At first glance one might think that an idea has gone well beyond the dreaming stage when it gets embraced and adopted by the United States military. Indeed, if there is any branch of the US government that has its feet firmly planted on the ground and deals with the unalterable brute facts of empirical reality, it would be the Department of Defense. Nevertheless, there exists within the defense department an agency that is explicitly devoted to far-out concepts and to "thinking outside of the box." And that is DARPA, the Defense Advanced Research Projects Agency.

DARPA (originally ARPA) was created by President Dwight D. Eisenhower on February 7, 1958, in response to the Soviet Union's launching of *Sputnik* in October 1957. *Sputnik* was the first artificial Earth satellite in history, and its sudden appearance overhead had taken everyone by surprise. A later book about the satellite was called *Sputnik: The Shock of the Century*. The satellite had four whip antennas attached to its outer surface and these transmitted a series of beeps that were soon picked up by amateur radio operators and by the BBC. The beeps, which sounded a little like cricket chirps, were in turn broadcast around the world and sent shivers down many a spine. The sounds were scary, eerie, chilling. It was as if there were Russian space aliens up there orbiting the Earth.

The satellite led to the "Sputnik Crisis" and to the Space Race between the USA and the USSR. *Sputnik* also prompted a reorientation of United States military priorities, for the government vowed never again to be overtaken by another country's strategic technological innovations. Thus the birth of an

"Advanced Research Projects Agency." The agency would be so very advanced that the Eisenhower administration authorized it to execute "research and development projects to expand the frontiers of technology and science far beyond immediate military requirements."

In 2010, one of those "far beyond" projects was DARPA's establishment, together with the NASA Ames Research Center, in Mountain View, California, of the 100 Year Starship Project. The program was first announced informally by Pete Worden, NASA Ames director, in a public interview sponsored by the Long Now Foundation in San Francisco, on October 12, 2010.

Toward the end of the interview, Worden said:

> We've just started a project with DARPA, it's called a Hundred-Year Starship Project. The idea is we're going to try to set up little mini-grants, and set up a program that will begin to get us to invest in the technologies of a Hundred-Year Starship. I think it's going to be done and we're hoping to inveigle certain billionaires to form a Hundred-Year Starship fund. So that's going to start, absolutely, we're going to build the project. DARPA sent me a million bucks, and I put a hundred grand of our [NASA Ames] money into it."

At this point the moderator, Peter Schwartz, asked: "The Starship, are you going to be on board?" To which Worden answered: "I'm probably not going to live that long."

This interchange gave rise to the impression that that the goal of the project was for DARPA and NASA Ames to build an actual, physical, working, starship that would take 100 years to construct. Which was not really the idea at all. In a later *New York Times* story about the scheme ("Offering Funds, U.S. Agency Dreams of Sending Humans to the Stars"), author Dennis Overbye wrote: "The Darpa plan has generated buzz as well as befuddlement" among space scientists. The befuddlement had been caused by the fact that, as a phrase, a "Hundred-Year Starship Project" was ambiguous. It could mean any one of three things. One, DARPA and NASA Ames are going to build a starship. Two, the process of building the starship would take a hundred years. Three, once launched, the starship would take a hundred years to get to its stellar destination. (Conceivably, it could also mean all of those things together.)

But DARPA was not in the spaceship business (nor for that matter was NASA Ames), and was not intending to build a starship. David Neyland, who was the director of DARPA's Tactical Technology Office, and who had come up with idea in the first place, explained that its goal was not necessarily to build a starship, despite what Pete Worden had said in his Long Now Foundation informal remarks. "One hundred years from now," Neyland said later, "there will be capabilities coming out of this that benefits us in the Department of Defense and the civilian sector, but also give us the capabilities of building the starship if we choose to do so." Which was a big "if."

What these agencies were actually doing was clarified somewhat on October 28, 2010, by an official DARPA news release titled: "DARPA/NASA Seek to Inspire Multigenerational Research and Development." This was an unfortunate title given the prevalence, among interstellar travel enthusiasts, of talk about multigenerational *spacecraft*. DARPA was emphatically not talking about *that*, however, but was rather referring to a research program that spanned generations. The press release stated, in part:

> The 100-Year Starship study will examine the business model to develop and mature a technology portfolio enabling long-distance manned space flight a century from now. This goal will require sustained investments of intellectual and financial capital from a variety of sources. The year-long study aims to develop a construct that will incentivize and facilitate private co-investment to ensure continuity of the lengthy technological time horizon needed.

Translated into English, what all of this rhetoric meant was that DARPA and NASA had jointly realized that nobody in their right mind formulated plans and undertook projects on anything like the 100-year time horizon that they thought was needed to design, build, outfit, and launch a crewed interstellar vehicle. So they wanted to seed-fund some private organization to do so, and for an essentially backdoor reason: namely to reap whatever possible spinoff technologies might accrue from such an endeavor. The press release continued: "DARPA also anticipates that the advancements achieved by such technologies will have substantial relevance to Department of Defense (DoD) mission areas including propulsion, energy storage, biology/life support, computing, structures, navigation, and others."

In other words, DARPA was proposing to fund a private starship research group essentially in order to gain whatever militarily useful technologies might be created as a byproduct of the group's effort to plan a trip to the stars. Given the fact that DARPA was a part of the defense department, that motivation made some sense. What was *assumed* by project, however, and not proven, was that crewed interstellar travel was possible and practical within the next 100 years. Therefore, despite the fact that startup funding was being provided by a branch of the defense department, the whole idea of interstellar travel still remained essentially just a dream. It would be the business of the private organization to show otherwise, to demonstrate how and that it could all be done.

But even as a dream, the idea of the US Department of Defense getting behind the concept of crewed interstellar travel was significant. It had done so once previously, with Project Orion in the 1950s and 1960s, before finally abandoning it. How long the 100 Year Starship Project would ultimately last was anybody's guess.

# # #

Pete Worden's initial announcement had been made in the fall of 2010. The next step forward in what would turn out to be the somewhat peculiar, wandering, and ill-defined trajectory of the 100 Year Starship project took place the following year, on the evidently symbolically meaningful date of 1/11/11, when DARPA and NASA Ames sponsored a strategic planning workshop at the Cavallo Point Lodge in Sausalito, California. This was a rather grand and historic collection of structures that had started out in 1850 as the Lime Point Military Reservation, later became Fort Baker, and in 2005, after considerable renovations and improvements, had been converted into a set of facilities that offered "a luxurious hotel experience with breathtaking views, a green ethos, and exceptional amenities."

Located at the foot of the Golden Gate Bridge, with views across San Francisco Bay, this was a posh resort where rooms started at $500 per night and where anything, including star travel, seemed possible. Here there now arrived a collection of about 30 experts from various disciplines, among them space, genomics, science fiction, futurism, and "other." The assorted visionaries included Mae Jemison, MD and former astronaut; Jill Tarter, extraterrestrial intelligence researcher; Marc Millis, interstellar travel advocate and head of the Tau Zero Foundation; Joe Haldeman, science fiction writer; and J. Craig Venter, formerly of the Human Genome Project, and now a speaker and consultant in various subject areas. Leader of the group discussions was Peter Diamandis, multitalented serial founder of several organizations including the X Prize Foundation and the Rocket Racing League; co-founder of the Zero Gravity Foundation; the Singularity University; and the International Space University in France; plus ancillary related ventures. Pete Worden, the NASA Ames director, was also present, as were three people from DARPA, including, of course, David Neyland.

The group was charged with systematically addressing what the organizers considered to be the three fundamental questions of interstellar travel: the "Why, What, and How." Meaning, in the first and most important case, why go to the stars?

This was a curious question for a project aiming at interstellar travel to be asking, given that the effort had already been funded by DARPA and NASA Ames to the tune of more than $1 million. It seemed to get matters exactly backwards, for in the normal course of events you started with a known objective and then tried to find funding for it. But in this case the organizers had put up the money to begin with and were now wondering: Why should we go to the stars in the first place?

But the real reason for asking was that the conference attendees were in the process of establishing guidelines for setting up the private organization that, when formed, would gather additional financing and commitments toward the goal of designing and building a starship. And in order to make a plausible case for anyone to support such a grandiose, otherworldly, and unlikely scheme, the relevant individuals or institutions would need a clear statement of the rationale and intent behind it.

But what this body of eminent advice-givers actually came up with was nothing really new, nor was it even very convincing. According to Marc Millis, who participated in the discussions, "it became obvious that human survival, via expansion into space, was a key motivation." Another conference attendee, Henry Kloor, chief science advisor to the X Prize Foundation, agreed: "If we don't eventually spread out – I'm not saying tomorrow or even a hundred years – but if we don't get off the planet it is inevitable that we would go extinct, just like the dinosaurs. Either a natural or unnatural event will occur that will wipe us out."

But this was a remarkably weak rationale for planning a starship, for two reasons. One, it was the same thing that interstellar proponents had been saying for years. Even Tsiolkovsky had pointed out that "one cannot remain in the cradle forever," one had to get off the Earth. And two, leaving the Earth per se was not a justification for going specifically *to the stars*. As Marc Millis had said, what was needed to avoid extinction in the face of a planetary disaster was "expansion into space," or as Henry Kloor had put it, to "get off the planet." The question of destination then remained open, and was hitherto unanswered.

The "What" question was next on the agenda, specifically, "What does the organization need to do to fulfill those motivations?" The attendees responded by providing a laundry list of objectives to be achieved within a given number of years. Among them were: In five years, prove that other habitable worlds exist; create a world view of hope; and produce, or cause to be produced, a blockbuster movie that would generate $500 million in receipts. Farther off: Land humans on Mars and communicate at faster-than-light speeds.

Perhaps the most novel concept to arise from these rambling discussions was voiced by the genomics expert, Craig Venter. To avoid all the problems associated with the process of sending live humans across light years of space, instead send only human genomes, and let them gestate, germinate, and bloom into adult humans upon arrival. Since no human beings would be at the target star to give birth to, raise, and bring the potential humans into educated adulthood, all this would evidently be done by an army of robots and fleets of advanced artificial intelligence systems placed at the scene beforehand. All of this presupposed a technology that did not exist as yet, and might never.

Finally, the "How?" question, which was not, unfortunately, "How do we design and build a starship?" but rather: "How can an organization be created and how can it achieve such milestones?" There was no single answer to this, given the sweeping nature of what was being asked. The only point that everyone seemed to agree upon was that the proposed organization should not be the government itself, nor any part of it.

Even DARPA's David Neyland believed this. "Looking at history," he said, "most significant exploration, like crossing oceans or continents for the first time, was sponsored by patrons or groups outside of government." What was needed, he claimed, was a method that gets the project "out of the government, and make sure it is an energized and self-sustaining enterprise."

# # #

The workshop ended without the DARPA or NASA Ames planners saying what the next step was or when it would be taken. But on May 5, 2011, DARPA issued a formal "Request for Information (RFI) 100 Year Starship Study." This was their bureaucratic way of soliciting written proposals for creating the organization that would begin the process of seeking further outside funding, and then the even more arduous, indeed awesome task of exploratory planning for the actual starship. The RFI promised the winning organization an initial start-up amount "not to exceed several hundred thousand dollars." Responses were due just a month later, on June 3, 2011. The RFI also noted that "100 Year Starship" and "100YSS" were the property of the United States and would be trademarked.

The winner was supposed to have been announced on the puzzling but also somehow symbolically meaningful date of 11/11/11. But DARPA missed that mystic mark, and, after sifting through dozens of proposals, made its decision known in late December. The winning team would be headed up by Mae Jemison, the former astronaut and the first Black woman of any nationality go into space.

Even among a group of people renowned for overachievement – NASA astronauts – Mae Jemison was a standout. Born in Chicago to working-class parents, she was an early fan of *Star Trek* and identified especially with the character Uhura, played by Black actress Nichelle Nichols. Jemison aimed for a career in science but was also interested in the arts, particularly dance, and studied ballet beginning at age 9. She was academically precocious and entered Stanford University at the age of 16. There she earned two degrees: a BS in chemical engineering and a BA in African and African-American studies.

Jemison then entered Cornell Medical School, in New York City, where she earned her MD degree in 1981. While in New York she also took dance classes at the Alvin Ailey American Dance Theater. After her medical internship and some years in private practice, Jemison joined the Peace Corps and served in Liberia and Sierra Leone.

In 1985, she returned to the States and while in private practice in Los Angeles first applied to NASA's astronaut training program. She was accepted into the fold in 1987, one out of a total applicant pool of 2,000. She became a Science Mission Specialist and flew on the Space Shuttle *Endeavour* in 1992, spending almost eight days in Earth orbit.

When she heard about the 100 Year Starship project, Mae Jemison, who had a pronounced "can-do" attitude toward taking on new challenges, considered herself as well-qualified as anyone else to lead it. And so she submitted a proposal to the organizers called "An Inclusive, Audacious Journey Transforms Life Here on Earth and Beyond." The document has not been made public and so its exact contents are unknown, but it turned out to be the winning entry. The award money in the amount of $500,000 went to the Dorothy Jemison Foundation for Excellence, which Jemison had founded in honor of her mother. In her new role as leader of the starship project she would be aided and abetted by two partner organizations: Icarus Interstellar, a space advocacy group headed by physicist Richard Obousy, and the Foundation for Enterprise Development, which fostered advanced research. Together, the three partner groups would jointly pursue their audacious dream. DARPA, meanwhile, stepped away from the 100YSS project forevermore. "We don't intend to carry it forward," David Neyland said. "DARPA hands the keys over to this entity and we wish them well."

One of the central tasks now facing Jemison's nascent organization was that of getting the general public behind the idea of going to the stars. In its October 2010 news release, the DARPA coordinator for the project, Paul Eremenko, was quoted as saying that "The 100-Year Starship study is about more than building a spacecraft or any one existing technology. We endeavor to excite several generations to commit to the research and development of breakthrough technologies and cross-cutting innovations ... to advance the goal of long-distance space travel, but also to benefit mankind."

But public attitudes toward spaceflight in general had waxed and waned over the years. A 2014 report by the National Research Council showed that interest was highest during the Apollo program's Moon landings, and also during the Space Shuttle era, and was lowest during the second half of the decade 2000–2009. During that time, the report said, "the level of public interest in space exploration is modest relative to that in other public policy issues." People were far more interested in abortion rights, climate change, and other environmental issues than in spaceflight, which they saw as hugely expensive while yielding only marginal payoffs. While there was always a small coterie of hard-core, dyed-in-the-wool interstellar flight zealots, the wider public remained cold to the idea.

# # #

Mae Jemison's response to this was to whip up public interest in long-distance spaceflight by sponsoring annual 100 Year Starship Public Symposiums. These gatherings, which would be open to the general public, would explore all aspects of interstellar flight in depth. The first of these meetings was the 2012 100YSS Public Symposium, billed as "Transition to Transformation: The Journey Begins," held at the Hyatt Regency Houston over three days in September 2012. By the time the event convened, the symposium had collected about a dozen backers, including Pfizer, Inc., United Airlines, Rice University, and the Houston Museum of Natural Science.

Into the conference center there now trekked about 300 paid registrants. They included scientists, futurists, venture capitalists, college professors, students, attorneys, journalists, *Star Trek* fans, *Star Trek* stars (LeVar Burton and Nichelle Nichols), science fiction writers, artists, ministers, and assorted others, most of whom were mentally geared up toward advancing the ambitious goal of hatching a scheme by which human beings could leave the Earth behind, travel to the stars, and establish a new Earth 2.0 in another solar system, all within the next 100 years.

Honorary Chair of the 2012 100YSS Public Symposium was none other than former president Bill Clinton, who issued a statement saying: "This important effort helps advance the knowledge and technologies required to explore space, all while generating the necessary tools that enhance our quality of life on earth."

At the conference itself, many of the speakers devoted their talks to the subject of interstellar spacecraft propulsion. This made sense because everything about interstellar flight depended on having a powerful, reliable, and advanced propulsion system. Without such a thing, it made little sense to discuss any other aspect of star travel. And so at the conference the first and by far the most important set of presentations was devoted to so-called "Time–Distance Solutions," bent on answering the question, how do we get there really fast?

And so on Friday, September 14, Richard Obousy, the head of Icarus Interstellar (one of the three partners of the 100YSS consortium), gave a talk entitled: "Starship by Design: An Exploration of Cutting-Edge Interstellar Technologies." Obousy was well-suited to the task: he had a PhD in physics from Baylor University in Texas, and his doctoral dissertation, about extra dimensions of spacetime, had included a chapter on faster-than-light travel. The primary difficulty with interstellar travel, he now explained, was simply the distance to the stars. Even the "nearest" star, Proxima Centauri, is incredibly far away: 4.246 light years from Earth, or 25,300,000,000,000 miles.

In order to cross that enormous distance, Obousy proposed several exotic propulsion schemes, such as pi-meson propulsion, the use of Kaluza–Klein geometries, string theory, and miscellaneous others. In fact, Obousy was a staunch proponent of exploring the use of *all possible fields* from *all possible dimensions*, whether or not such fields or dimensions were known to be real as opposed to being merely theoretical entities. This sounded all too much like sheer fantasy, but that was often true of interstellar travel theory and speculation.

Adding to that impression was a similar talk by Obousy's colleague from Baylor, Gerald B. Cleaver: "Spacecraft Propulsion via Chiral Fermion Pair Production from Parallel Electric and Magnetic Fields." But the fact was that most of the various propulsion schemes proposed by Obousy, Cleaver, and others were conjectural: they were merely *ideas*, creatures of the imagination, possibilities, not realities. Further, the schemes that involved faster-than-light travel had the additional and substantial drawback that they violated a known, accepted, and fundamental law of nature: the fact that the speed of light – 186,000 miles per second – is an absolute speed limit in the physical universe.

For most scientists, this is accepted as an unbreakable law, and a blunt fact of reality. Except, that is, by interstellar travel zealots. It is an idiosyncrasy of interstellar travel theory, and theorists, that many of them are not daunted in the least by the thought of traveling at, or even faster than, the speed of light, in so-called "superluminal" or FTL (faster-than-light) spacecraft.

But the fact that such propulsion systems were speculative in the extreme, coupled with the fact that some of them verged on the physically impossible, only emphasized just how far away from actual realization interstellar travel really was.

It was a surprise, then, that amid all of this theorizing, one speaker at the 2012 100YSS Public Symposium claimed to have some actual, working hardware in a NASA laboratory where he was doing real, hands-on experiments on a faster-than-light spacecraft propulsion system. This was Harold G. "Sonny" White.

# # #

For a guy who has spent much of his professional life researching the possibilities of faster-than-light transportation systems, Harold G. "Sonny" White is a remarkably down-to-earth, nuts-and-bolts individual. He holds degrees in mechanical engineering from the University of South Alabama and Wichita State University, and a doctorate in plasma physics from Rice University. White grew up in Virginia just outside Washington, DC, and as a kid spent many hours at the Smithsonian's National Air and Space Museum, ogling planes, rockets, and other spacecraft, including the Apollo 11 command module *Columbia*. He had been thus bitten by the space bug early on in life and started working for NASA in the year 2000. While there he was manager of the Space Shuttle's remote manipulator arm division, and in 2006 he won NASA's Exceptional Achievement Medal for his work on the project. He later became leader of the Advanced Propulsion Team at the NASA Engineering Directorate of the Johnson Space Center in Houston.

The most advanced propulsion system Sonny White investigated was called the Alcubierre Drive. Within the ranks of interstellar travel theorists, this was a famous, almost legendary concept because it would make faster-than-light speed spacecraft possible and yet it was conceptually just so simple. The basic idea was that instead of moving the spacecraft from place to place by means of rocket propellant, you instead altered the fabric of spacetime itself in such a manner that the distant star in effect came to you.

This idea had initially been proposed by Miguel Alcubierre, a Mexican theoretical physicist, who in the early 1990s was studying for his doctorate in physics at the University of Wales, Cardiff. In 1994 he published in the peer-reviewed scientific journal *Classical and Quantum Gravity* a paper entitled "The Warp Drive: Hyperfast Travel Within General Relativity." While the body of the piece is chock full of equations unintelligible to non-specialists, the paper's opening abstract explains the essence of idea in clear, plain English.

> It is shown how, within the framework of general relativity and without the introduction of wormholes, it is possible to modify spacetime in a way that allows a spaceship to travel with an arbitrarily large speed. By a purely local expansion of spacetime behind the spaceship and an opposite contraction in front of it, motion faster than the speed of light as seen by observers outside the disturbed region is possible. The resulting distortion is reminiscent of the 'warp drive' of science fiction.

Later, in the text, the author said, as if to be absolutely unambiguous about it, that when all the necessary conditions required for the warp drive were satisfied, "The spaceship will then be able to travel much faster than the speed of light." *Much* faster!

There was, of course, a problem. "Just as it happens with wormholes, exotic matter will be needed in order to generate a distortion like the one described here," and, "exotic matter is forbidden classically." This did not doom the idea, however, because "it is well known that quantum field theory permits the existence of regions with negative energy densities in some special circumstances. ... The need of exotic matter therefore doesn't necessarily eliminate the possibility of using a spacetime distortion like the one described above for hyper-fast interstellar travel."

At some point during the course of his advanced propulsion research, Sonny White became intrigued with this quite radical concept. And in 2006, White had an idea for reducing the Alcubierre drive to practice; that is, by subjecting it to experimental laboratory test that would reveal the existence of an actual warp in spacetime. It would be only a tiny warp "bubble," less than half an inch in diameter, a volume in which spacetime would be perturbed, or "warped," by as little as one part in 10 million. White ran a laboratory at the Johnson Space Center called "Eagleworks" where he was putting together the hardware necessary to perform the space warp experiment.

At the 2012 100YSS Public Symposium in Houston, White gave a talk entitled "Warp Field Mechanics 102" in which he described his basic setup to a roomful of rapt listeners. "It's very, very modest," he said, "a microscopic instance of this phenomenon, nothing that you would try and bolt to a spacecraft by any stretch."

He projected pictures of his lab bench, the top of which was a pegboard onto which the various test devices were mounted and secured. At one end was a laser beam generator, and at the other end a mirror. Between the two was the detection and measurement apparatus. When in operation a bright red laser beam passed through the center of a capacitor ring that was about the size and shape of a wire loop that a child might swing through an arc to create soap bubbles.

"We have a ceramic capacitor ring that we charge up to many thousands of volts, to implement a potential energy that blue-shifts that frame relative to where the capacitor is located," he said. "What the field equations predict is that the presence of that potential energy and boost will create a spherical perturbation. So that's what we're trying to measure, to see if we can generate that change in optical properties in that little spherical region."

If, after the capacitor ring was energized, the two sides of the spherical space encapsulated by it were optically different, the disparity would be registered by a specially built field interferometer, a device that could detect and measure minute variations in electromagnetic wavelengths. That optical difference, supposedly, would be evidence of a tiny deformation of space – a space warp.

White had done a number of computer simulations that showed what the resulting spacetime distortion would look like. Pictured was a flat mesh that looked much like a home window screen stretched out horizontally: that was the fabric of spacetime. Resting on the center of the screen was a small, blue cylindrical object that was tapered at both ends: the spaceship. When the capacitor ring was energized, a deep depression would form in front of the spaceship: the contraction of spacetime ahead of the ship, pulling it forward. Behind it there was an ascending area of the meshwork that was of equal vertical magnitude to the size of the depression ahead of the ship: the expansion of spacetime behind the ship, pushing it forward. The simultaneous contraction and expansion of spacetime at the ship's two ends would move it forward at ultrafast speeds.

That, anyway, was the theory.

# # #

But apparently it was *only* a theory. White had given an earlier version of his Houston talk at a previous 100YSS symposium ("Focused Attention on Future Scientific Challenges"), held in September 2011, in Orlando, Florida. This was before DARPA had announced who would be eventually running the program, and the meeting was perhaps in the nature of an audition or dress rehearsal of what was to come, since two of the winners, Mae Jemison and Richard Obousy, participated in the Orlando meetings (as had also Jill Tarter, who was an omnipresent figure at space gatherings of all kinds). In Orlando White had given a prior talk, "Warp Field Mechanics 101," and it is notable that in neither presentation did White claim that he had actually detected or produced the one-centimeter spacetime bubble he talked about. He had not observed the optical difference that would be the physical, experimental evidence of a space warp. He spoke of his experiment as "hopefully an existence proof" of the phenomenon, and also hoped that there would be something real based on the deformation of spacetime "maybe within my lifetime." But there was nothing real yet. If anything, the lack of any observed effect cast doubt upon the existence and feasibility of the alleged phenomenon.

White himself conceded that there were some technical obstacles to realizing the warp drive in actual practice. "Spacetime is really, really stiff," he said at the 2012 symposium. "It's pricey real estate," and you'd need the mass of planet Jupiter to produce the space-warp effect on the scale of a full-size, working starship. He hoped to get around this, he said, by "manipulating the stiffness, reducing the stiffness" of spacetime, but how he was supposed to accomplish this amazing feat remained an open question.

Despite both the theoretical and practical problems associated with the space warp, White nonetheless exhibited at the 2012 100YSS meeting a computer-generated image of what a warp-driven starship could look like. It had been computationally generated by Mark Rademaker after Matthew Jeffries's 1964 concept design for the original *Star Trek* warp-drive spacecraft. Still, there are plenty of pictures of things that don't exist or wouldn't work, such as da Vinci's drawings of an ornithopter.

By the time of the next 100YSS public symposium ("Pathway to the Stars"), in 2013, the warp drive's conceptual inventor, Miguel Alcubierre, had himself given up on his very own theory. "It's a nice idea," he told a writer for *Popular Science*. "I like it because I wrote it myself. But it has a series of limitations that I've seen through the years, and I don't see how to fix them."

# # #

After its 2012 public symposium the 100 Year Starship held three additional public gatherings: in 2013 and 2014, in Houston, and one in 2015, in Santa Clara, California. There was no public symposium in the years 2016 through 2022. At that point it looked like the 100 Year Starship project had suddenly dwindled to just four years of actual existence. It hadn't, though, inasmuch as another one took place in 2023, in Nairobi, Kenya.

Much had already changed even by 2013, however. For one, in that year Richard Obousy had resigned from the 100 Year Starship, and had also stepped down as president of Icarus Interstellar (though he remained on as a director). The reason was that he had founded a new company, CitizenShipper. According to its Web page (citizenshipper.com), the company was "a two-sided marketplace for hard to-ship items such as dogs, cats, motorcycles, boats, cars, and more. CitizenShipper puts you in touch with experienced, background-verified, and user-rated dog transporters. A quality experience – quick, safe, and affordable!"

Obousy also explained why he made this shocking about-face turn from interstellar travel to this quite different and novel transportation concept. "I noticed that a lot of people drive trucks, especially in Texas, and often the truck beds are empty. I also recall gas prices getting pretty high around the years 2007, and as a graduate student I wasn't making much money. I initially started thinking how good it would be if I could make some extra money on the side using empty space in my vehicle to transport things for people. Then I started thinking, why limit it to just me? Why not create a website that services the entire country and connects people with spare cargo space with items that need shipping?"

So, Richard Obousy had solved one out of two problems of getting matter from one place to another: pets across the country, yes; human beings to the stars, no.

For Mae Jemison and the 100 Year Starship, there was also a brief change, from public symposium format to a type of event that took on a decidedly New Age aura. The year 2017 happened to be the 25th anniversary of Mae Jemison's flight aboard the Space Shuttle *Endeavour* in September 1992. To commemorate that occasion, Jemison announced a "LOOK UP Global Special Event." A news release on the 100YSS website dated

September 15, 2017, said: "LOOK UP over the next year will connect people worldwide, from all walks of life, culminating on a single day in August 2018 when everyone will be asked to LOOK UP and share what they see and their thoughts, hopes, fears, dreams, and ideas for the best path forward. LOOK UP is a day, 24 hours, we acknowledge our oneness as Earthlings and concurrently our right to be part of this greater universe." Whatever that meant.

The plan was for participants to share their thoughts, impressions, emotions, and photographs on a special, new website, lookuponesky.org (now defunct). The site launched on the day of the September 15 announcement, and then on the appointed 24-hour day of August 18, 2018, people could upload their images to it and became part of a unique "Sky Tapestry."

This actually happened. The experience was a sort of global, online astronomical be-in. What all of it had to do with advancing the prospects of interstellar travel, however, was left to the imagination.

3 THREE ICONS OF STAR TRAVEL

In the literature and lore of interstellar travel, three unique designs for spacecraft crop up again and again in the story: the Bernal sphere, the Bussard Interstellar Ramjet, and Project Daedalus. Each of the designs was proposed for a different specific purpose, but the first of them was so central to the idea of human star travel that it was also incorporated into the second of the three. And two of the schemes used the same means of propulsion: controlled nuclear fusion. None of these planned structures was ever actually built: they exist only on paper, in people's minds, in illustrations, and in computers. They are strictly creatures of the intellect, not real things, and might never be.

Further, this being the realm of interstellar travel theory, it is also true that none of the designs obeyed the general principles of standard engineering practice. In their own respective ways each concept was a blend of unrealistic assumptions about what was possible or practical in the indefinite future. Aspects of the schemes also reflected the view that since an object had a name, it also had existence. This was the problem of the correspondence-correlate: from the fact a name existed it did not follow that anything in external reality corresponded to it. For example, many of the exotic propulsion systems proposed by interstellar travel advocates are known to exist in name only: the warp drive, tachyon propulsion, the EM drive. But the repeated occurrence of these names in the literature of inter-stellar travel gives them a spurious credibility, a sort of implicit, shadow, twilight, almost presumptive existence. But granting them credence would be misplaced: they are still abstractions, or, in the language of Daniel Deudney in his critique of space expansionism, *Dark Skies*, they are "imaginaries," not realities.

Chronologically, the first of the three icons of star travel was the Bernal sphere, proposed by J. D. Bernal in 1929. Taken by itself, the Bernal sphere is not usually considered to be a starship, and indeed as initially proposed, it was not: Bernal intended his sphere first and foremost as a space habitat, to be built in space and located in orbit around the Earth. Later, however, Bernal advanced the further notion that the habitat could also serve as the centerpiece of a multigenerational vehicle that could carry a large human population across interstellar distances. Such a vehicle, which was also known as a "space ark," or "world ship," would ultimately emerge as one of the foundational concepts of crewed interstellar spaceflight. Since a world ship is essentially a Bernal sphere combined with a propulsion system, Bernal may be credited with having proposed what would be an integral, indeed indispensable part of a multigenerational interstellar vehicle. In addition, a Bernal sphere is also the kernel notion underlying the equally iconic notion of "human colonies in space" as proposed by Gerard K. O'Neill, and which was briefly famous during the 1970s. This is true despite the fact that O'Neill's colonies were generally cylinders (although he also proposed spherical colonies), and also differed from Bernal spheres in other important ways.

John Desmond Bernal was an Irish scientist who pioneered the use of X-ray crystallography, a method of imaging the fine structure of biological molecules. It was this technique that crystallographer Rosalind Franklin, working with James Watson and Francis Crick, employed to determine the double-helical configuration of DNA. According to Freeman Dyson, Bernal "began looking at the crystal structure of nucleic acids with X-rays long before DNA became fashionable."

Bernal was a graduate of Cambridge University and later became a professor of physics at Birkbeck College, University of London, and was a Fellow of the Royal Society. He had a checkered career, was a member of the communist party of Great Britain, and while in the Royal Navy took part in the

D-Day invasion of Normandy. He was also a prolific writer of books, both popular and technical, and is best known for his 1929 tract, memorably entitled *The World, the Flesh, and the Devil: An Enquiry into the Future of the Three Enemies of the Rational Soul*.

Bernal's idea was to explore the ways in which science could perfect or at least improve upon the material and emotional conditions of human life. The "world" referred to the physical deficiencies commonly faced by many on our home planet: scarcity, hunger, overcrowding, poverty, fire, flood, and so on. Here Bernal wanted to "revolutionize the whole of human life and to turn the balance definitely for man against the gross natural forces." The "flesh" represented bodily infirmities such as disease, aging, and death. His solution was to use medical science to augment human capacities and abilities, yielding, at the end a "mechanized man" endowed with a healthy life span of hundreds of years. And the "devil" symbolized the darker forces within human nature. Bernal's remedy for such ills was somewhat nebulous but seemed to lie in the realm of a technologically etherealized mass consciousness – whatever that might be.

Bernal's prescription for improving the material conditions of life on Earth was to leave it behind, and to build a permanent home for a civilization in space.

> Imagine a spherical shell ten miles or so in diameter, made of the lightest materials and mostly hollow …. Owing to the absence of gravitation its construction would not be an engineering feat of any magnitude. The source of the material out of which this would be made would only be in small part drawn from the earth; for the great bulk of the structure would be made out of the substance of one or more smaller asteroids, rings of Saturn or other planetary detritus. … The globe would fulfil all the functions by which our earth manages to support life. In default of a gravitational field it has, perforce, to keep its atmosphere and the greater portion of its life inside; but as all its nourishment comes in the form of energy through its outer

surface it would be forced to resemble on the whole an enormously complicated single-celled plant.

It is difficult to see why Bernal regarded living in this structure – the Bernal sphere – to be an improvement upon ordinary and everyday life on Earth. And indeed he himself allowed that "Criticism might be made on the ground that life in a globe, say of twenty or thirty thousand inhabitants would be extremely dull, and that the diversity of scene, of animals and plants and historical associations which exist even in the smallest and most isolated country on earth would be lacking." Additionally, since the sphere does not rotate there is no gravity, and since there is no bad weather inside the sphere, houses would become superfluous. "The major part of the lives of the inhabitants of the globe would be spent in the free space which would occupy the greater portion of the center of the globe."

This free-floating, drifting lifestyle would take some getting used to:

Objects would be endowed with a peculiar levity. We should have to devise ways of holding them in place other than by putting them down; liquids and powders would at first cause great complications. An attempt to put down a cup of tea would result in the cup descending and the tea remaining as a vibrating globule in the air.

Such problems were solved easily enough by the first astronauts, crew members of the Space Shuttle, and those living aboard the International Space Station. Others were not, however, particularly the problem of avoiding or rectifying the hazards to human health resulting from prolonged weightlessness: impaired functioning of the immune system; muscle atrophy; loss of bone mass; loss of blood volume; alterations of visual acuity and the shape of the eye; and other such degenerative effects. In light of these and other factors, when Gerard K. O'Neill advanced his designs for space colonies, he provided the cylinders with artificial gravity. "Gravity is easy to find in space," he wrote in *The High Frontier: Human Colonies in Space*. "Rotation can provide it. On the inside of a hollow rotating vessel the gravity can be made the same as on Earth."

In spite of the somewhat fantastic nature of his scheme, Bernal was realistic enough to recognize that the sphere would be constantly subject to meteorite damage. "The presence of meteoric matter in the solar system moving at high speeds in eccentric orbits would be the most formidable danger in space travelling and space inhabitation."

This was proven to be all too true by NASA's Space Shuttle program, because by the end of it more than 100 shuttle windows had been replaced after impacts with space debris. Some objects were as small as the fleck of paint that cracked the front window of STS-7 (the second *Challenger* mission) in 1983. After a while there had been such a volley of debris impacts that the shuttles, once they reached orbit, were intentionally flown tail-first to minimize the effects of collisions. The damage inflicted by any given impact only increased with the speed of the spacecraft, and starships traveling at relativistic speeds through the interstellar medium would be particularly susceptible to such impacts.

In *The World, the Flesh, and the Devil*, Bernal predicted that attempts at interstellar flight were all but inevitable. "A globe which was merely a satisfactory way of continuing life indefinitely would barely be more than a reproduction of existing conditions in a more restricted sphere." In other words, a Bernal sphere in Earth orbit was by itself fairly pointless. (Daniel Deudney wryly referred to O'Neill cylinders as "canned Earths.") The solution was to go elsewhere, meaning, of course, to the stars. Bernal continued:

> Interstellar distances are so large that high velocities, approaching those of light, would be necessary; and though high velocities would be easy to attain – it being merely a matter of allowing acceleration to accumulate – they would expose the space vessels to very serious dangers, particularly from dispersed meteoric bodies. ... Even with such velocities journeys would have to last for hundreds and thousands of years.

Because it was so far-out, speculative, and fantastic, none of Bernal's projections exactly had the ring of truth. The idea of *allowing acceleration to accumulate*, for example, sounded as if the process happened almost automatically, without human agency or other intervention. This was armchair theorizing, not the product of sound science, calculation, or practical engineering. Still, for all of its fantasies, the Bernal sphere stands as a milestone proposal among the various intellectual constructs underpinning hopes of traveling to the stars.

# # #

The concept of journeys lasting for "hundreds and thousands of years" was abhorrent to Robert W. Bussard, the architect of what is without question the most charismatic and alluring starship of them all, the Bussard Interstellar Ramjet. If there were a rock star of star travel, it would be this. Paul Gilster, long-time interstellar travel advocate and administrator of the website *Centauri Dreams*, has written: "The interstellar ramjet conceived by Robert Bussard may have launched more physics careers than any other propulsion concept." And with good reason. The ramjet was a system by which a starship would scoop up free hydrogen molecules from the interstellar medium, feed them into a fusion engine, and convert them into particles of thrust. The starship would thus be gathering "fuel for free" as it sailed through space. As such, the device was sort of a cosmic perpetual motion machine – almost impossible but not quite – and therefore it attracted a lot of attention among starship zealots. Carl Sagan, in *Cosmos*, was quite enthusiastic about it, despite his claiming that the ramjet would have an intake scoop that would be "hundreds of kilometers across." (That would turn out to be a gross underestimate.) Sagan also said of the fusion engine that "This is engineering on a scale so far unprecedented on Earth. We are talking of engines the size of small worlds."

Earlier in his career, in 1962, when Sagan was a member of the department of genetics at Stanford University Medical Center, he had written a paper about human travel to the stars and said that: "I believe that interstellar spaceflight at relativistic velocities to the farthest reaches of our Galaxy is a feasible objective for humanity. ... It is concluded that with nuclear staging, fusion reactors, and the Bussard interstellar ramjet, no fundamental energetic problems exist for relativistic interstellar spaceflight." Even for Sagan, that was optimism taken to extremes. (Also in this same paper, Sagan had suggested that, "the Earth has been visited by various galactic civilizations many times (possibly $\sim 10^4$) during geological time.")

The conceptualizer of the device, Robert W. Bussard, was an American physicist with a Princeton PhD who had worked for much of his life on nuclear fusion energy research. Nuclear fusion is of course the reaction that powers the stars, and is the energy source for thermonuclear bomb blasts. Still, controlled nuclear fusion was notoriously difficult to achieve in the laboratory, and required the use of immense machinery, extremely high temperatures in the range of millions of degrees, enormous amounts of energy, and the expenditure of countless billions of dollars.

The story is told (improbable as it might seem) that Bussard got the idea for a fusion-powered ramjet at breakfast one day while he was working at the Los Alamos Scientific Laboratory. He'd ordered a tortilla with a scrambled egg inside it, and supposedly that cylindrical shape and its filling made him think of a ram starship (Figure 3.1). In 1960, at any rate, he submitted a paper that has come to be one of the most frequently cited and discussed publications in the interstellar literature: "Galactic Matter and Interstellar Flight." He submitted it to the journal *Astronautica Acta*, whose editor at the time was Theodore von Kármán, an astrophysicist who was known as a tough critic. As audacious as Bussard's scheme was, von Kármán nevertheless accepted it for publication.

Although the bulk of the paper is highly technical, Bussard stated its core argument quite simply. The problem with conventional chemical rockets, he said, was that it would take them too long to reach the stars.

Figure 3.1 Bussard interstellar ramjet. (Wikipedia Commons)

Since considerable inert mass must be expelled and maximum attainable velocities are small relative to the speed of light, an optimum ... interstellar rocket powered by conventional nuclear energy sources will require flight times of hundreds of years to reach even the nearest stars.

Bussard emphatically did not accept this as reasonable, although he conceded that others did.

Others have considered as a possible solution the concept of interstellar travel involving flights of hundreds, perhaps thousands of years, with whole civilizations in microcosm rising and falling while in flight between planetary worlds. If we wish to avoid this aesthetically unattractive picture, yet cling to hope for interstellar travel, we must find a way to overcome the inadequate energy-source objection cited above.

It is the purpose of this paper to discuss one method of doing this, by abandoning the interstellar rocket entirely, turning to the concept of an interstellar vehicle which does not carry any of the nuclear fuel or propellant mass needed for propulsion, but makes use of the matter spread diffusely throughout our galaxy for these purposes.

Bussard then went on to note that the interstellar void, although it was commonly regarded as "empty space," is in reality filled with several different types of matter. Distributed throughout the galaxy, for example, are neutral hydrogen atoms, which exist in a density of one or two atoms per cubic centimeter (a space about the size of a pea). He also noted that such atoms were not spread evenly throughout the cosmos, but were congregated in arrays of clouds and in various filamentary structures.

In addition to neutral hydrogen, the interstellar medium also contains ionized hydrogen atoms, which are atoms that have lost an electron, leaving only a proton. Protons are positively charged particles, and they strongly repel one another. In interstellar space, the ions would be collected by a scoop at the front of the ramjet vehicle, and would then be fused together by the engine, which would release energy in the process.

[The ions] are deflected by an electric or magnetic field which causes them to arrive at a focal point ... At the focal point these ions are led into a fusion reactor of unspecified (indeed, unknown) type, made to react and generate power which is then fed back into the fusion process through a similarly unspecified conversion device, to the increase of their kinetic energy and momentum, with consequent reaction on and acceleration of the vehicle.

Simple as that.

# # #

Bussard's scheme involved a lot of unknowns, unspecifieds, and crucial points to be filled in later. But these are inescapable elements of interstellar travel speculation and theorizing, and so Bussard's paper caused a mild sensation among the small and insular interstellar flight community of scholars. Many of the scientists familiar with the physics and operation of the ramjet were concerned about the feasibility of the intake scoop. In his original paper, Bussard himself proposed "an ion collector radius of nearly 60 km [37 miles]," which he conceded was "very large by ordinary standards." But subsequent researchers found that this value was much too small to provide the speeds that Bussard envisioned; others noted that the scoop would be subject to physical stresses, strains, and forces ignored by Bussard. A 1969 paper by John Ford Fishback ("Relativistic interstellar space-flight") pointed out that at high speeds the incoming hydrogen ions would produce a substantial amount of drag on the vehicle and would also subject the scoop field magnets to mechanical stresses. To accommodate these forces, Fishback proposed some changes to the structure, operation, and size of the intake scoop.

Later, a 1978 journal article by Thomas Heppenheimer ("On the Infeasibility of Interstellar Ramjets") came to an even more pessimistic conclusion: if the interstellar medium was "optically thick," then the intake scoop would have to be half a light year across, whereas if it were optically thin, then the drag produced by a phenomenon called *bremsstrahlung* radiation (a braking effect) would exceed the total amount of energy gained. In other words, the drag force generated by the passage of the large scoop through the sea of particles that made up the interstellar medium would exceed the thrust generated by the engine, making the entire device self-defeating.

Other authors proposed various further alterations and additions to the ramjet, or suggested hybrid versions. There was the "Ram Augmented Interstellar Ramjet," offered by Alan Bond writing in the *Journal of the British Interplanetary Society* in 1974. It added a plasma accelerator into the mix. A year later, in 1975, Dan Whitmire advanced the idea of a "Catalytic Nuclear Ramjet" in which carbon was carried onboard the ship to catalyze the nuclear fusion reaction. In 1977, Whitmire and Albert Jackson proposed the "Laser Powered Interstellar Ramjet" in which laser energy beamed up from Earth was used to help push the spacecraft along. Through the years to come, the number of variants to Bussard's original concept continued to blossom and grow.

Everyone involved in the discussion of the Bussard Interstellar Ramjet in its many incarnations agreed that there was a problem with the device, but no two authors seemed to agree on what the problem was nor precisely how to solve it. Paul Gilster almost put the whole concept to rest by saying that, in the end, Bussard's ramjet scoop "may turn out to have more applicability in braking than in acceleration." But new versions and analyses continued to appear.

A recent (2022) study by Peter Schattschneider and Albert A. Jackson ("The Fishback Ramjet Revisited") returned to Fishback's redesigned intake scoop and used computer simulations to evaluate the feasibility of the device. "Our results indicate that there are severe engineering problems with the magnetic collection process," the authors said. To be effective, they argued, the intake scoop would have to be some 2,000 kilometers (1,200 miles) across. That was far less than Heppenheimer's record-breaking half-a-light year scoop, but it was still unrealistically large. But there was even worse news ahead than that: the authors described the need for "absurdly long magnetic coils" in order to keep the incoming molecules within the cone. The value they suggested for the length of the coils was 150 million kilometers (93 million miles), which coincidentally just happens to be the distance from the Earth to the Sun.

The final conclusion of the authors was that "It is very unlikely that even Kardashev Civilizations of type II might build ramjets." Kardashev civilizations were units of a classification scheme that ranked hypothetical civilizations according to the amount of energy they used. Civilizations of type I used the amount of energy that reached a planet from its parent star. Earth is therefore classified as a type I Kardashev civilization. Civilizations of type II have harnessed all of the energy emitted by its own star, perhaps by enclosing its entire solar system within a Dyson Sphere, a megastructure that encompasses a star and all of its orbiting planets. In saying that even a type II Kardashev Civilization would be "very unlikely" to build an interstellar ramjet is therefore equivalent to saying that its construction would be impossible for the type I civilization of mere planet Earth.

So, what had started out as an apparently elegant and attractive solution to the problem of interstellar propulsion ended up as a Brobdingnagian collection of ever-escalating and complexifying ramjet concepts. Unsurprisingly, the only truly successful application of the Bussard Interstellar Ramjet was in the service of science fiction, as several authors including Poul Anderson in his book, *Tau Zero*, embraced the device for fictional space travel purposes. Numerous artists have also depicted the ramjet in action, and in full color, traveling at full speed on its fantastic, headlong, semi-perpetual-motion blitzkrieg through the interstellar deep.

As for the Bussard Interstellar Ramjet itself, it remains firmly embedded within the realm of the hypothetical and the imaginary. An actual working example is about as far out of reach to us as flying to the Moon was to Lucian of Samosata.

# # #

The third of the iconic starships was the most comprehensively, thoughtfully, and professionally designed of the three: the Project Dedalus spacecraft. This was a creation of the British Interplanetary Society, which itself had an illustrious history. It had been founded in 1933 by Philip E. Cleator, in Liverpool. Cleator was just 25 years old at the time and was an avid space fan, having been inspired by science fiction books and movies, as well as by the American Interplanetary Society (later the American Rocket Society), founded in 1930 by a group of science fiction writers. In 1936 Cleator published a book of his own about space flight: *Rockets Through Space: The Dawn of Interplanetary Travel*. Practically from its inception, the British Interplanetary Society (BIS) became one of the world's most important and influential space advocacy groups. Its stated purpose was to advance and support the science of astronautics, and to promote the idea and practice of space exploration.

The society published its own scientific journal, *Journal of the British Interplanetary Society* (*JBIS*), and, later, its own books. This was a decidedly serious group, composed of educated scientists and engineers, as well as members of the general public. It was in no sense a bunch of *Star-Trek*-bedazzled space cadets.

The nonprofit group eventually moved its headquarters to the Vauxhall area of London, from which members of the society hatched various schemes for the exploration of space. In the late 1930s, for example, the society proposed a manned expedition to the Moon by the use of multistage, solid-fuel rockets. Later still, after the Apollo Moon landings and the surge of ideas for expeditions to Mars, the society, in spite of its eponymous "interplanetary" focus, widened its scope and turned its attention to the stars.

Members of the BIS met regularly at The Mason's Arms pub on Battersea Park Road in London. In 1973, over drinks, 11 members, including Alan Bond, who was a Rolls-Royce rocket designer, plus Anthony Martin, Bob Parkinson, and others, conceived of a spacecraft for traveling to Barnard's Star, a red dwarf some 5.9 light years from Earth. Astronomer Peter Van de Kamp, director of Sproul Observatory at Swarthmore College, had published data that appeared to suggest a wobble in the star, which he theorized was caused by the gravitational pull of a planet that was approximately 1.6 times the size of Jupiter orbiting the star. Later observations by others showed that Van de Kamp's claim was incorrect, and said that the wobble was actually produced by a flaw in the telescope's lensing system, not by a planet.

That made no difference to the Project Daedalus engineers, however, for they had already established a set of ground rules to govern their starship design process, one of which allowed for a change of destination before launch or even while in flight. Their first guiding principle was that the spacecraft had to use current or near-future technology. That is, the craft could not be a collection of hypothetical or imaginary features that had no prospect of real-world existence: no warp drives, tachyons, or wormholes need apply. The second rule was that the craft had to be designed to reach its destination within a human lifetime, or in the space of about 50 years. And the third was that the vehicle had to be one that could aim for a variety of different target stars.

The group's planning process took some five years, and the design that the members came up with incorporated a number of novel, radical, and imaginative ideas, although the spacecraft they envisioned had a significant drawback from the standpoint of human interstellar travel: the vehicle would carry no crew, and would be "manned" only by robots. Even so, their concept was ambitious.

For one thing, their projected spacecraft was so large that it would have to be assembled in Earth orbit. In addition, it would also be built from materials derived from one or more other planets within the solar system, or from their moons, or both. Indeed, the project presupposed a human civilization that had already expanded far beyond the Earth and had occupied and industrialized other bodies of the solar system. In their account of the design, *Project Daedalus: The Final Report on the BIS Starship Study*, published in the *Journal of the British Interplanetary Society* in 1978, the authors stated that their spacecraft "fits naturally into the context of a solar system-wide society making intelligent use of its resources, rather than a heroic effort on the part of a planet-bound civilization."

In fact, the project was heroic for any civilization. This was planning on the grand scale.

# # #

Much of the spacecraft's design was dictated by the distance to Barnard's Star, 5.9 light years from Earth, and the time allotted to get there, which the BIS team put at 50 years. Further, they calculated that crossing that distance in the given time frame would require a cruising speed of about 12 percent the speed of light. This meant that conventional propulsion methods such as those used in chemical rocketry would be insufficient to the task. And for that reason the planners turned to nuclear fusion, a method that did not exist at the time but was one that they expected would be developed in the near future.

The design that finally emerged called for a two-stage, nuclear-pulsed rocket. The first stage would be fired for two full years, taking the spacecraft to 7.1 percent the speed of light. At that point, the first stage would be shut down and jettisoned. The second stage would then be fired for 1.8 years, after which the craft would cruise for the next 46 years at 12 percent the speed of light before arriving in the vicinity of Barnard's Star, which it would glide past without stopping or braking.

Since the spacecraft, having been built in Earth orbit, would not travel through the atmosphere, it was not at all streamlined. Both stages resembled a collection of large spherical fuel tanks stuck together at a common center. This arrangement looked like a bunch of coconuts topped by a bunch of grapes. The first stage consisted of six large tanks containing the bulk of the propellant; the second stage consisted of four smaller tanks containing the remainder of the propellant. The entire vehicle was some 600 feet from top to bottom, not quite twice the height of NASA's Saturn V Moon rocket.

A novel feature of the spacecraft was that the designers had provided for two separate devices for avoiding or deflecting collisions with any interstellar matter likely to be encountered enroute. Immediately atop the leading second stage was a large beryllium disk – 7 mm thick by 64 meters (210 feet) in diameter and weighing 50 tons – that would shield the spacecraft from incoming particles. But the disk was not fully up to the task intended for it, since it did not span the full width of the first stage whose immense fuel spheres bulged out far beyond the width of the disk.

In addition, therefore, flying in open space 200 kilometers (125 miles) ahead of the spacecraft would be an artificial cloud of particles that would itself absorb and capture some of the incoming hazardous materials. The cloud was to be generated by a fleet of support vehicles called "dust bugs." But exactly how this apparatus worked, and how successful it would be in practice, and against incoming objects of what size and mass, were open questions.

But assuming that all these systems worked, then upon reaching the target star, the Daedalus craft would release into space 18 artificially intelligent autonomous probes which would explore the planet that theoretically orbited Barnard's Star, and then radio information about their discoveries back to Earth.

The conceptual and functional core of the spacecraft was its propulsion system, which in this case was a fusion reactor embodying the hoped-for invention of controlled nuclear fusion. The development of controlled nuclear fusion technology is one of the most ambitious, complex, and difficult tasks in the history of science and technology.

Nuclear fusion is the process in which two or more atomic nuclei, such as hydrogen (H) nuclei, are fused to form a molecule of a heavier type, such as helium (He). Vast amounts of energy are released in the process, and the reaction also generates an array of subatomic particles (neutrons or protons), plus intense heat, gamma rays, and X-rays. Since hydrogen nuclei (protons) are positively charged particles, they strongly repel one another and it takes massive amounts of energy, in the form of temperatures of millions of degrees, to get them to fuse together. The high temperatures required are why nuclear fusion reactions are called *thermonuclear* reactions (and why hydrogen bombs are called thermonuclear bombs).

There are two main types of nuclear fusion reactions. The first is the sudden and explosive type, as takes place in hydrogen bombs. One of the problems in developing such a bomb was creating enough heat to get the reaction started and propagated without it fizzling out. Los Alamos scientist Enrico Fermi once proposed to Edward Teller the idea of igniting the fusion reaction by using a conventional fission bomb – an atomic bomb – as a sort of igniter or spark plug.

That was in fact the method by which the world's first thermonuclear device was detonated. Los Alamos scientists working on the Manhattan Project had nicknames for their volatile creations. The very first atomic bomb – the one detonated at Trinity Site on July 16, 1945 – was simply called "the gadget." Later, when many of those same scientists had constructed the first thermonuclear device, they called it "the sausage," after its shape. As built, the sausage had a core of lithium-6 deuteride that was to be triggered by the explosion of a plutonium fission bomb lying lengthwise through the center of it.

The amount of energy given off by a thermonuclear bomb blast was graphically illustrated by the test of the sausage, which took place on the island Elugelab, part of Eniwetok Atoll in the Pacific, in 1952. Elugelab was a 40-acre diamond-shaped land mass that was part of a coral atoll located 11 degrees above the equator about 3,000 miles southwest of Hawaii. It was uninhabited except by birds and was covered by tropical grasses, wildflowers, shrubs, and small trees known as *Messerschmidia*.

The test itself was code-named "Ivy Mike," and took place on Elugelab on November 1, 1952. It is an understatement to say that the device was successful, for within a matter of seconds after the countdown reached zero, what had once been a peaceful tropical islet had vanished into thin air. In its place was a mile-wide crater filled with water 160-feet deep. Gordon Dean, the chairman of the Atomic Energy Commission, responsible for the test, reported the results by saying, somewhat gleefully, "The island of Elugelab is missing!"

The amount of energy generated by explosive, or "uncontrolled" nuclear fusion is what makes controlled nuclear fusion so difficult to achieve, for controlled nuclear fusion is the same explosive process as occurs in a nuclear bomb, only stretched out or extended over a substantial time span. It might be thought of as a persistent or continuous explosion, a thermonuclear bomb blast in the form of an extended smear of energy across time.

Such a controlled, sustained, and contained thermonuclear reaction is difficult to achieve technologically. The first problem is getting it started. The second problem is keeping the reaction going, and at a steady rate, without its fizzling out. And the third problem is containing the energy liberated by the fusion reaction, which would exist in the form of a superheated plasma. A superheated plasma cannot be contained or confined by a physical structure because the plasma exists at temperatures in the range of millions of degrees, which would instantly melt any sort of physical housing. The plasma would have to be confined instead by magnetic, electric, or other fields or forces. But it was just this controlled fusion process, with all of its challenges, questions, and unknowns, that the Project Daedalus designers were hoping would power their spacecraft and propel it to the stars.

Specifically, the BIS engineers pinned their hopes on a novel thermonuclear propulsion scheme proposed by Friedwardt Winterberg of the University of Nevada Desert Research Institute, in a 1972 paper, "Thermonuclear Micro-Bomb Rocket Propulsion." In it, the author proposed "an advanced nuclear rocket propulsion system based on a chain of exploding thermonuclear micro-bombs ignited in front of a concave magnetic reflector open at one side." These micro-bomb fuel pellets would each contain a mixture of helium-3 and deuterium, and ignition would be produced by "an intense relativistic electron beam."

Winterberg intended the system for use in interplanetary travel within the solar system, predicting that it "will make possible flights to Mars in less than a week, to the planet Jupiter in less than a month and to the planet Pluto at the edge of our solar system in less than a year." The Project Daedalus designers, however, sought to adapt this system for use in interstellar travel within the Galaxy.

As the Daedalus engineers conceived of it, the fuel would exist in the form of small pellets that would be injected into the reaction chamber by means of an electromagnetic gun, and then detonated at the rate of 250 pellets per second. This was a system in which nuclear energy would be released by a continuous series of mini-explosions as opposed to a stretched-out individual explosion. The hot plasma stream that resulted from those explosions would be directed out by a magnetic nozzle, thus generating the rocket's thrust. For a trip that would last four full years, an immense volume of pellets would be required as fuel. The designers calculated that in the actual rocket there would be 50,000 tons of pellets, divided up into 46,000 tons carried in the large first stage, and 4,000 tons in the second stage (50,000 tons is about the total weight of 15 Saturn V rockets).

Of course there were some problems with all this. As the authors of the Daedalus Final Report themselves said, "The major problem with using the D3He [deuterium/helium-3] reaction is the fact that 3He is virtually non-existent on Earth." Their solution was to go to Jupiter and mine it for its plentiful helium-3 supplies: "atmospheric processing in Jupiter should be able to provide ample He for many vehicles," they said. The mining operation would be accomplished by a fleet of 128 large robotic factories over a period of some 20 years.

# # #

Any assessment of Project Daedalus has to start with the fact that even in the event that all of it worked properly, the spacecraft would not take humans to the stars, the reason being that the ship was designed from the beginning to be uncrewed. It might be thought that this lack could remedied simply by adding a crewed component, a Bernal sphere, for example, equipped with all the relevant necessities for maintaining a human founding population across a 50-year time span. Such an addition, however, would so radically alter the overall makeup of the craft that all the original design parameters would be fundamentally changed.

The project was an attempt at a proof of concept, showing that a flight to the stars within a reasonably short time frame was possible in principle. But even as such, the design had a number of flaws. One was that the design contained many component parts, the failure of any one of which could be fatal to the spacecraft. A malfunction of the artificial cloud of particle deflectors maintained by the "dust bugs," for example, could let space debris impact and possibly destroy the first stage fuel tanks, portions of which were not shielded by the secondary and backup beryllium disk. The dust bugs themselves could fail. The fusion reaction system, likewise, was made up of many parts, any of which could malfunction: the electronic gun particle injector; the smooth and steady supply of 250 particles per second; the magnetic nozzle directing the craft; the electronic beam confinement system. Every element would have to work perfectly and continuously for the space of four years, whereas none of these components was yet a tested and proven working technology.

Further, the claim that Project Daedalus would use "near term" technology was not in fact adhered to by the designers inasmuch as they assumed an already humanized and industrialized solar system. How far we are from that state of affairs may be gauged by observing how much has been done to get humans back to the Moon, or off to Mars, or to Jupiter, or anyplace else in the solar system (other than Earth orbit) in the 50 years since the last Apollo mission in 1972, which is to say, very little. In those 50 years we have seen robotic vehicles on Mars dig small trenches into the surface, and a robotic helicopter that has flown over the Martian plains. We have crashed a rocket into an asteroid and altered its trajectory, and we have brought a small quantity of asteroidal material back to Earth. But humans themselves have made no further expeditions out into space, much less have we industrialized the entire solar system. The BIS dream of "a solar system-wide society making intelligent use of its resources" therefore remains, like much else, just a dream.

The Project Daedalus design study was published in 1978. More than 30 years later, in 2010, a group representing a new generation of scientists associated with the advocacy organization Tau Zero Foundation collaborating with the British Interplanetary Society jointly undertook a critical examination of the original Daedalus engineering concept and its underlying assumptions. The collaborative group found many shortcomings with the design. For one thing, they thought that mining helium-3 from

Jupiter was problematic given the enormous gravity well that would have to be overcome to lift materials off the planet. "A source of fuel other than the gas giant Jupiter should be considered," they said.

The group was also skeptical of the Daedalus fusion propulsion system. "This entire operation is a technical challenge in itself even for a reactor stationed on earth," they wrote. Making it all work in a spacecraft was an even trickier business. Indeed, the technology of controlled nuclear fusion was as yet unproven. The achievement of controlled nuclear fusion was a goal that receded ever away from you as you approached it, and was perpetually some 20 years off in the future.

In view of the Project Daedalus deficiencies, the group of collaborators decided to produce a revised design of their own. They described their planning process in a paper called "Project Icarus: Son of Daedalus," and published it in the May 2010 issue of the *Journal of the British Interplanetary Society*. The members did not present detailed new design specifications, but simply advanced general ideas for what they described as an "unmanned probe that is capable of delivering useful scientific data about the target star, associated planetary bodies, solar environment and the interstellar medium."

They also issued a number of guidelines, principles, and key watchwords "to ensure that all design solutions are appropriate." Using capital letters to emphasize the tenor and importance of their concerns, they said that "The final design must be a CREDIBLE proposal and not based on speculative physics. It must be a PRACTICAL design. It must be derived using accepted natural laws and using SCIENTIFIC methods which are supported by experiments." These and other design criteria were unusually strict for the sometimes otherwise overly imaginative, too-clever-by-half approach to the process of designing interstellar spacecraft.

It was no surprise, then, that over the course of the redesign process the group members failed to produce a workable, credible, and practical design for an Icarus spacecraft. In 2020, fully 10 years after the start of the design process, three members of the collaborative group issued a new paper, "Project Icarus: Designing a Fusion Powered Interstellar Probe," and published it in the scientific journal *Acta Futura*. "Despite the nearly 100 internal reports and publications, the effort to design an 'Icarus' craft was stalling by 2013," they wrote, "Team members were unable to agree on the best option for a fusion system, or even the best fuel."

The group came up with hosts of new named spacecraft designs, among them: Firefly, Ghost, Resolution (later renamed Endeavour), Zeus, Leviathan, Pathfinder, and Starfinder, but did not manage to find one that they could all agree on. The group also identified a number of new fusion reaction system concepts, including Z-pinch fusion and plasma jet magneto inertial fusion, and also considered the use of several different fuels: deuterium–tritium (D–T), deuterium–helium 3 (D^3–He), deuterium–deuterium (D–D), and ultra dense deuterium (UDD).

In the end there were far too many competing designs rather than too few, and the group's inability to settle on any one combination of spacecraft design, reaction system, and fuel effectively doomed Project Icarus to the dustbin of interstellar spacecraft history. There has been little further activity on the project since 2019. Although they were intended to fly to the stars, projects Daedalus and Icarus both wound up going nowhere.

4 PROJECT ORION

The Project Orion spacecraft is by common consent the craziest interstellar flight concept ever devised. Ironically, it was also the spacecraft design that received the widest support by scientists, the military and other branches of the US government, as well as by private industry. It was as if all of these people had collectively lost their minds.

The basic idea was utterly simple and so intuitively obvious that it could be understood by a child. This was a craft whose propulsion system was built upon the Newtonian principle of action and reaction. The central notion was that of placing a bomb under a rocket and then detonating it to loft the rocket up and away – exactly the same process as putting a firecracker under a tin can and watching it blow sky high. To keep it going up, of course, a series of bombs detonated in sequence would be required. And so the Orion rocket would be propelled through space by a stream of bombs, in fact nuclear bombs, exploding one after another behind it, thereby continuously accelerating the craft.

That was the project's key concept, and as such it was simultaneously perfect and insane. The advantage of the setup was that where the rocket could go, and how fast it could get there, were limited only by the number of available bombs and how powerful they were. With enough bombs and enough explosions, you could go anywhere, including to the stars. The disadvantage was that nuclear bombs are among the most destructive devices ever created by human beings. Putting them in close proximity to a spacecraft, possibly one containing people, maybe hundreds or even thousands of them, was just nuts.

This was in essence the village idiot's idea of space travel, the comic book version of interstellar flight. It was tailor-made for the movies, or, perhaps even better, for use as a mass transport vehicle in the Super Mario Bros video game. But as lunatic as it sounded, the surprise was that in 1955 two Los Alamos nuclear scientists had advanced the scheme as a sane, sober, and realistic spacecraft design possibility. Equally surprising was the fact that practically as soon as they had put the idea on paper, the concept was taken up by other scientists who were not part of the Los Alamos group, after which it was also actively embraced by ARPA (predecessor of DARPA), the Atomic Energy Commission, the US Air Force and the Department of Defense, and finally even by NASA. Between the years 1957 and 1965, still other scientists working at General Atomic (later General Atomics), a nuclear energy and defense corporation based in southern California, produced more than 260 technical reports, many of them classified as "Secret – Restricted data," about the design's feasibility, its technical specifications, principles of operation, practicality, possible uses, and potential hazards, which were considerable. The total amount of funding devoted to the project was more than $11 million, including about $1 million from General Atomic itself.

Further, some of the Orion's intellectual backers were among the luminaries of science: Freeman Dyson, a mathematical physicist at Princeton's Institute for Advanced Study; Ted Taylor, who as a Los Alamos project scientist was credited with having developed the smallest, the most powerful, and the most efficient nuclear fission weapons ever tested by the USA; and Edward Teller, the so-called "father of the H-bomb." Even so levelheaded and calm a personage as Charles Lindbergh was also onboard with the idea. Others, however, ridiculed the project. Richard Feynman called Orion "the pie in the sky." But a scheme as crazy as this one was bound to be controversial.

Still, the US government and General Atomic fostered research on Project Orion and poured money into it for seven consecutive years. And then, in January 1965, they pulled the plug and it all ended with a thud. As an ongoing research effort, the project was now dead. The idea of a nuclear-bomb-propelled spacecraft, however, continued to thrive ever afterward and to exert its strange appeal.

# # #

There had been a few anticipations of the Orion concept through history. The Chinese black-powder rocket was itself an early example, but it had been intended only as a toy or weapon, and not as a vehicle for travel purposes. A more modern design for a "rocket airplane" propelled by pulsed explosions was advanced by the Russian Nikolai Kibaltchitch in 1881. Kibaltchitch was an amateur bomb-maker and revolutionary who took part in the assassination of Czar Alexander II (he supplied the bombs). According to the rocket historian Willy Ley,

> Kibaltchitch had visualized his rocket airplane as a platform with a hole in the center. Mounted above this hole was a cylindrical explosion chamber into which "candles" of compressed powder were to be fed. The apparatus was to rise, vertically at first, then the chamber was to be tilted for horizontal motion; the speed was to be regulated either by the dimensions of the powder "candles" used or by their number.

Kibaltchitch died in a Russian prison shortly after proposing the scheme.

In 1940, science fiction author Robert Heinlein described an atom-bomb-propelled spacecraft even before the atomic bomb had actually been invented. He depicted it, somewhat skeptically, in his story, "Blowups Happen," in which a character asks:

> You mean atomic fuel for interplanetary flight? The problem's pretty well exhausted. ... Of course, you could mount the bomb in a ship, and figure out some jury rig to convert its radiant output into push, but where does that get you? One bomb, one ship – and twenty years of mining in Little America has only produced enough pitchblende to make one bomb.

As a realistic scientific concept, the atom-bomb-propelled spacecraft had to await the development of a genuine, working atomic bomb. The first such device had been tested successfully at Trinity Site in the Alamogordo desert in 1945. Ten years later, in 1955, two Los Alamos project scientists, Stanislaw Ulam and Cornelius Everett wrote a paper (initially marked Secret, later declassified), entitled "On a Method of Propulsion of Projectiles by Means of External Nuclear Explosions."

Ulam was one of the original Manhattan Project scientists. He had helped develop both the fission bomb and later the thermonuclear fusion bomb, or H-bomb, of which, along with Edward Teller, he was the co-inventor. Everett was a colleague of Ulam's who worked on the problem of fusion bomb ignition. In their joint paper they wrote:

> The scheme proposed in the present report involves the use of a series of expendable reactors (fission bombs) ejected and detonated at a considerable distance from the vehicle, which liberate the required energy in an external "motor" consisting essentially of empty space. The critical question about such a method concerns its ability to draw on the real reserves of nuclear power liberated at bomb temperatures without smashing or melting the vehicle.

Without smashing or melting the vehicle, yes.

The genesis of the concept, they said, went as far back as 1946, when Ulam came up with the idea just the year after the fission bomb had been successfully tested. (Another account, in George Dyson's book *Project Orion*, says that Ulam first thought of the idea "the day after Trinity.") As for the problem of melting the rocket, a possible solution, the authors suggested, was to create a strong magnetic shield that would deflect the ionized propellant atoms away from the base of the rocket while nevertheless transferring momentum to the ship itself. The craft they had in mind was a small one, "saucer-shaped, of diameter about 10 meters, sufficient at any rate to intercept all or most of the exploding propellent."

The "propellent" (as they spelled it) was not the bomb itself, nor the heat and light generated by the explosion, but rather a mass of material "consisting of water or some plastic, which will be heated by the bomb, and which will propel the vehicle during its subsequent explosive expansion."

> The bombs are ejected at something like one second intervals from the base of the rocket and are detonated at a distance of some 50 meters from the base. Synchronized with this, disk-shaped masses of propellent are ejected in such a way that the rocket-propellent distance is about 10 meters at the instant the exploding bomb hits it. The propellent is raised to high temperature, and, in expanding, transmits momentum to the vehicle.

That, anyway, was the bare-bones concept, with many details left to be worked out later. The authors acknowledged some dangers with the proposed design, one being the possibility of predetonation of the remaining bombs by neutron flux from those already exploded outside the vehicle. Here also they suggested that shielding would overcome the difficulty. The issue of stability, which they saw as "probably a major problem," had not been addressed. Ulam and Everett also admitted that "wastefulness of fissionable material" was a disadvantage of the scheme – a shortcoming that has never bothered later proponents of the atom-bomb-powered spacecraft.

At Los Alamos, physicist Ted Taylor read the Ulam–Everett paper. Taylor had an uneasy relationship with nuclear weaponry. In grad school studying nuclear physics at Berkeley, he declared himself a firm opponent of their use: the bombings of Hiroshima and Nagasaki had appalled him. On the other hand, he also thought that if managed intelligently, the threat posed by a massive nuclear arsenal could end the prospect of war once and for all.

It was perhaps not so odd, then, that his first job after leaving grad school in 1949 was at the Los Alamos National Laboratory, home of the US nuclear bomb program. Taylor was totally wowed by the Ulam–Everett paper. If ever there was an ideal use for nuclear bombs, he thought, it was this. "The first flight of that thing doing its full mission would be the most spectacular thing that humans had ever seen," he thought.

The launch of *Sputnik* in 1957 further galvanized his thinking. By that time Taylor had left Los Alamos and gone to work for General Atomic, a division of General Dynamics Corporation. General Atomic had been founded in 1955 by Frederic de Hoffmann, another Los Alamos expatriate. Hoffmann had been born in Vienna but came to the United States to study physics at Harvard. While at Los Alamos he'd worked with Edward Teller on the development of the hydrogen bomb, but left in 1955 to start General Atomic, a firm that would commercialize nuclear reactors for energy production by private industry, and to pursue other peaceful uses of nuclear technology.

When Ted Taylor arrived at General Atomic in 1956, one of the projects he worked on was to help design a nuclear reactor that would be so safe to operate that it could be used by private enterprise – "an ultra-safe research reactor," as Freddy de Hoffmann described it. That effort would culminate in the TRIGA reactor: "Training, Research, Isotopes, General Atomic," to produce short-lived isotopes for use in hospitals. The first TRIGA prototype became operational in 1958.

TRIGA was a practical device that became a worldwide success. The other project that Ted Taylor worked on at General Atomic was not obviously practical: Project Orion. Inspired first by the Ulam–Everett paper and then goaded further on by *Sputnik*, Taylor was motivated to come up with a design of his own. He described it in his report, *Note on the Possibility of Nuclear Propulsion of a Very Large Vehicle at Greater than Earth Escape Velocities*. It was dated November 3, 1957 (coincidentally the day that *Sputnik II* was launched), and furnished the design specifications for a vehicle weighing 4,000 tons, which carried 2,600 tons of bombs and propellant, and which was able to put a payload of 1,600 tons into Earth orbit. Taylor estimated that the craft could be built within six or seven years, at a cost of some $500,000,000. He named the vehicle Orion, after the Orion constellation; "I just picked the name out of the sky," he said.

Ted Taylor and Freddy de Hoffmann submitted the proposal to ARPA early in 1958 and sat back to await results.

# # #

The technical director of ARPA at that time was physicist Herbert York, who did not himself regard the Orion rocket as credible, but who in the throes of *Sputnik*-induced technological space mania was nevertheless willing to give the concept a chance. "I never thought it was feasible," he said later. "But that's OK, I thought it was interesting ... [and] even very low feasibility merited some attention."

The result was an Air Force contract, "Feasibility Study of a Nuclear Bomb Propelled Space Vehicle." Dated June 30, 1958, it authorized a grant to General Atomic in the amount of $999,750 for 13 months of engineering and design work on the project. "If the concept is feasible," the contract said, "it may be possible to propel a vehicle weighing several thousand tons to velocities several times earth escape velocities. Such a vehicle would represent a major advance in the field of space propulsion."

The document went on to describe the fundamental operating principles of the craft, noting that the system consisted of a circular pusher plate connected to the ship proper by a shock absorber mechanism. A series of nuclear bombs stored in the ship are fired below the pusher plate. The bomb when fired is surrounded by a propellant, which at detonation is vaporized into a plasma that strikes against the pusher plate, driving it forward into the shock absorbers, "which then deliver a structurally tolerable impulse to the ship."

The contract gave General Atomic what it needed to put together an engineering and design team that would specify the size, shape, and precise operational features of the craft. Ted Taylor would become the executive director of the project. He was well suited to the job: as a boy he had placed cherry bombs underneath 50-gallon drums and watched them fly into the air to a height of as much as 15 feet. And later, as a nuclear bomb designer at Los Alamos, he became a veritable whiz kid at inventing custom-made bombs for a variety of purposes.

Other members of the team, which at the start numbered only a dozen or so, included Jerry Astl, explosives expert; Brian Dunne, lead experimentalist; and Marshall Rosenbluth, a plasma physicist who would concentrate on the problem of pusher plate ablation.

And then there was Freeman Dyson. Dyson was arguably the most famous scientist ever to have been associated with Project Orion. The word "infamous" would be equally apt in his case, for Dyson, a mathematical physicist at the Institute for Advanced Study in Princeton, has been described as a "maverick," "iconoclast," "heretic," and an "eccentric," all with good reason. His life as a scientist had been marked by an extraordinary profusion of ideas, some of them quite far-out, on a wide range of diverse subjects. He even defended the concept of crazy ideas and crazy people. "Have you ever been to Cambridge University?" he once asked a journalist. "It's full of crazy people – oddballs, loners on the verge of doing something really tough and historic. Why shouldn't they be crazy? Nature is crazy. I would like to see more crazy people here at the Institute."

One of his own crazy ideas was proposing the human colonization of comets in the Oort cloud. "Comets, not planets, may be the major potential habitat of life in the solar system," he wrote in his essay "The Greening of the Galaxy." We may even ride them to the stars, he said, since comets may "provide way stations for the migration of life all over the galaxy."

Dyson, who was born in 1923 in Crowthorne, England, was an early bloomer. As a boy he had read Jules Verne's story *From the Earth to the Moon and a Trip Around It*, and decided, at age nine, to write a science fiction story of his own. It was called "Sir Phillip Roberts's Erolunar Collision," and was about the forthcoming collision of the asteroid Eros with the Moon. Sir Phillip and members of an astronomical society resolved to make a voyage to the Moon themselves in order to observe the impact firsthand. The young Dyson furnished some preliminary design specifications of the spacecraft that would take them there, saying that it "need not exceed in length 25 feet, and in diameter 10 feet; so the projectile, with rooms and telescope, need not exceed 32 ft. in length, and 15 ft. in diameter." Separately, he proposed a "celestial" spacecraft, apparently for interplanetary travel, which "need not exceed two miles in length."

Two miles in length! Already, at age nine, Freeman Dyson was thinking big. Given this, and his lifelong attachment to rockets ("I was always in love with spaceships," he said), plus his adult conceptual craziness, there was probably no one better suited to work on Project Orion than he.

Dyson was brought into the project in 1957 when Freddy de Hoffmann visited him at his office at the Institute for Advanced Study. Hoffmann made this an in-person visit because the subject was still too secret for a phone call. "He came here and he said, 'Look, you've got to come to GA,'" Dyson recalled later. "I said no, I have no intention of coming to GA . . . And he said no, you must come, we have something much bigger and much more exciting, and then he told me Ted had this wonderful scheme for getting around the solar system with bombs."

Dyson liked the idea from the start. "It sounded good," he recalled afterward. "It didn't frighten me. The immediate reaction from everybody is that it will blow the ship to pieces. I wasn't bothered by that. The thing made sense on a technical level. It sounded like what we'd all been waiting for. This was an alternative to chemical rocketry that could work."

In fact, Dyson liked the idea so much that he made a trip to General Atomic after all, arriving at first only as a consultant. He had known Ted Taylor since 1953; they had met at Cornell University, where Dyson was teaching and where Taylor was a student. General Atomic was located in La Jolla ("the jewel"), and Dyson would take brief leaves of absence from the institute and stay at the Hotel Del Charro, which was near the beach. Later, when he began fulltime work on the project, General Atomic provided him with a house in La Jolla with a swimming pool, a citrus orchard, and a spectacular view of the Pacific.

# # #

Table 4.1 Physical parameters of Project Orion vehicle concepts.

	Gross weight	Diameter	Height
Recoverable test vehicle	50–100 tons	40 feet	50 feet
Orbital test vehicle	880 tons	80 feet	120 feet
Interplanetary ship	4,000 tons	135 feet	200 feet
Advanced interplanetary ship	10,000 tons	185 feet	280 feet

Now came the business of actually designing the nuclear-bomb-propelled spacecraft. There was no shortage of designs. There was, to start with, the one that Ted Taylor had outlined in his 1958 proposal to ARPA: 4,000 tons gross weight, 2,600 tons of bombs and propellant, and a payload of 1,600 tons. Soon there were others: a surviving document from early on in the project shows four different vehicle concepts (Table 4.1).

Later, in September 1959, Dyson gave a talk describing an Orion spacecraft that could get a colony of several thousand people to Alpha Centauri in about 150 years. His son, George Dyson, explains, "Only one problem: it would require 25 million hydrogen bombs to get there – and 25 million more if you wanted to stop."

At one point there was even talk of a "Super Orion," with a mass of 8 million tons and a diameter of 400 meters (1,312 feet, which was almost a quarter-mile across). Still, departing from these extremes, the scientists settled on a ship that more or less approximated the original as proposed by Ted Taylor. At the very time that Project Orion began, General Atomic was building a circular library on its campus in La Jolla. The diameter of the library was 135 feet, and coincidentally or not this was also the diameter of the most popular Orion design concept.

"We intended to build a spaceship which would be simple, rugged, and capable of carrying large payloads cheaply all over the solar system," Freeman Dyson said. "Our slogan for the project was 'Saturn by 1970.'" He and many others wanted – and expected – to go there themselves personally.

But there were lots of things that could go wrong even with such a "simple and rugged" atom-bomb-powered spacecraft. First there was the problem of the integrity of the pusher plate across time, while being subjected to thousands of nuclear explosions in close succession. All of the many design concepts for the Orion spacecraft had in common a pusher plate that was large, curved, thick, and strong, and was made out of materials that would resist the effects of the blast. In one 1958 design, the pusher plate was a reinforced, 1,000-ton steel or aluminum disk that was four inches thick and 120 feet in diameter, and was shaped so that its mass distribution matched the momentum distribution of the explosion. In addition, the surface of the disk would be covered by a sacrificial ablative layer of material that would be vaporized away, thus protecting the disk from damage. A new layer would be applied by spray nozzles between each successive blast. Finally, the shock wave from each blast lasts for only a very short time – 1/3 millisecond – also minimizing the effects of each pulse. The highest heat from the explosion, in the range of 100,000 to 120,000 degrees, lasts a mere fraction of a second. But it was unknown what the effects on the pusher plate would be after several thousand such hits.

The pusher plate was almost one-third of an acre in area. Was it realistic to think that a new coating of ablative layer, a heavy oil spread by nozzles, could be applied smoothly and evenly over such an area within the short time interval available – one-third of a second or less – between blasts?

Then there was the problem of transporting the bombs from the storage area in the payload section of the ship to the space below the pusher plate where they would be fired. There were two theories about how to do that. One was a classic Rube-Goldberg-style masterpiece of complexity. In their report issued in September 1958, Orion project authors Don Mixson and Ron Prater wrote that: "The firing of about 2,600 bombs at the rate of 4 per second is a tremendous problem." As so too was their proposed solution:

This system consists essentially of a series of pairs of rails located around the periphery of the pusher and behind it. The bombs are launched from behind the pusher and, upon sliding down the last portion of the rails are ignited and given (by curved rails) just enough pitch motion that in flight they follow a curved path around the edge of the pusher to the point of desired detonation. . . . How to detonate the bombs at the correct time and place is another question that has not been seriously considered yet.

The scheme was an engineering nightmare and was abandoned soon enough.

The other method of bomb delivery was the "obvious" one: namely, drop them through an ejection tube that ran down through the center of the pusher plate and out into the open space below. A trap door flap would open up as the bomb charge approached the bottom end of the tube, and close immediately after the bomb fell through it. A short spray of anti-ablation oil would follow the closing of the trap door, after which the bomb would be fired.

There were of course problems with this. The bomb blast plasma would hit the pusher plate and drive it up into the shock absorber assembly, which would dampen the force of the blast and then bounce back down to receive the shock wave from the next blast. But what would happen if the next bomb in sequence was a dud? Then there would be no way of stopping the rebounding pusher plate from being shot off into space.

The other difficulty was that of getting all this to work reliably and seamlessly thousands of times, with bombs going through the trap door three to four times per second, and with the 4-inch-thick trap door opening and closing each time with exquisite precision. "The trap door was a formidable problem," Freeman Dyson said much later. "It has to open and close in a fraction of a second, under very rugged conditions with things going bang all the time. If the shutter jams shut then you're finished."

An ancillary problem was that of getting all the bombs lined up in proper sequence for delivery into the ejection tube. Since this was like getting Coke bottles from a vending machine, the designers sought help from the Coca-Cola Company. "One major source of information," Ted Taylor explained, "was the Coca-Cola Company, where we got detailed descriptions of how they set up mass production systems for Coke bottles and all kinds of mechanical stuff." The general idea was that the bombs (also called "pulse units"), numbering in the thousands (or, later, even millions), are stored in individual helical magazines, and descend, one by one, through a chute that releases the next one as the previous one is ejected.

The third essential component of the nuclear-bomb-propelled spacecraft was the shock absorber system: its function was to convert what amounted to a heavy sledgehammer blow to the ship into a tolerable, gentle shove. A document drafted toward the end of the project described a two-stage apparatus: "The first stage shock-absorber system, a series of concentric gas-filled tori, serves to reduce the initial pusher acceleration to approximately 100g's. An intermediate platform couples the first-stage shock absorbers to the second stage, piston-cylinder shock-absorbers, which in turn reduce the accelerations felt by the upper structure of the engine to a few g's or less."

Altogether, these three separate systems amounted to a lot of complicated machinery with multiple opportunities for things to go wrong at every point, and it was hard to know in advance if all of it would work as planned, again and again, over a long stretch of time. Ted Taylor likened the range of potential problems to "unspecified types of car trouble, if the car is too complicated."

# # #

One way of finding out whether such complicated systems would work as planned was by running a series of tests using small-scale models of the full-size ship. As it happened, when such tests were made in 1959, they showed quite graphically just how many "unspecified types of trouble" could occur.

A separate reason for doing small scale-model tests of the Orion concept was to prove to doubters that bombs need not destroy the vehicle. As Jerry Astl, the explosives expert on the Orion team explained it, they had to convince people of the project's feasibility:

> because every time I approach some higher-ups I hear the stereotype answer: we use bombs to blow things to pieces not to make them fly. So that was why there was so much emphasis put on making the damn thing fly, so they can show them it is not necessary to blow things to pieces with bombs, they can be used for good practical purposes.

It took a long string of trials to establish that point, however. The scale-model flight tests were held at Point Loma, a high cliff overlooking the Pacific just west of San Diego. The place had been outfitted with gun batteries during World War II, and still held remnants of emplacements that had walls six-feet thick. There were ammunition bunkers and other hefty structures around the area as well, making the site perfect for conducting experiments with small amounts of high explosives. There was also a 75-foot-tall tower that had been used for static tests of Atlas missiles and that could now be used to launch the mini-Orions.

The experimenters started out with a very small, primitive, home-made model that they jokingly referred to as "Spaceship Number One." It had an aluminum pusher plate topped by three nested, stainless steel mixing bowls. The model was suspended 10 inches above 60 grams of a common explosive known as C-4. When detonated, according to one observer, "A few coyotes and some deer got scared, a skunk in the nearby bush fired back, but 'Spaceship Number One' was unharmed." There is no public record of how far, if at all, the model flew.

The engineers then built a second model with a pusher plate that was just one foot in diameter. The pusher plate was aluminum, the bullet-shaped hull above it was made out of fiberglass, while the shock absorbers consisted of layers of polyurethane foam. The explosive charges used by this model were golf-ball sized, and after more than 50 shots there had been no ablation of the pusher plate, but by the end of the tests the shock absorber material had crumbled almost entirely to dust.

Still, these tests suffered from a serious limitation, which was that neither of the models had an ejection tube through which bombs traveled down and through the pusher plate, whereas this was arguably the most important and complex mechanism envisioned for the full-size Orion rocket. And so the Orion engineers built yet a third model, one that incorporated an ejection tube that could hold up to six bombs that could be fired in close succession.

This final model was one meter in diameter and again had a fiberglass hull. As George Dyson describes it in his authoritative book, *Project Orion*, "For the meter model, a series of high-explosive charges – grapefruit-sized balls of C-4, shaped by hand and cushioned by polystyrene foam inside coffee-can-sized canisters – were ejected through the middle of the pusher plate at quarter-second intervals from a central stack." When fully loaded the model weighed 270 pounds. They called it "Hot Rod." (Air Force officials, a bit dismissively, referred to it as "Putt Putt.")

It took a long time to get Hot Rod/Putt Putt to work properly. The first problem was that of determining the right amount of explosive to use for best results. Too small a quantity, and the thing wouldn't get off the ground. "We'd have a big roar of high explosive, but the damn thing would not move, it would just sit there bathed in this low-density gas," said experimenter Brian Dunne. *Kaboom! Kaboom! Kaboom!* Followed by silence.

Using too much explosive, on the other hand, did in fact "blow the thing to bits." This happened again and again. The thing would spring up through the launch tower, zoom up into the sky, and then – *Kaboom!* – it would explode in a puff of smoke and rain back down to Earth in a million pieces.

One time Ted Taylor invited mathematician Richard Courant to come out and observe the flight tests. Courant, who was an advisor to the project, was from Germany and spoke with a heavy German accent. So he came out one Saturday morning in the summer of 1959 and watched the test. The bomb went off, Hot Rod lifted up through the launch tower, and then – *Kaboom!* – disintegrated in front of his eyes like a firecracker.

"Zis is not nuts, Zis is supernuts," Richard Courant said.

# # #

Between the extremes of failure to launch on the one hand and blowing itself to pieces on the other, a long series of individual glitches, malfunctions, misfires, jams, and other snafus plagued the flight tests, which started in June 1959. In one test flight, at a height of five feet, the shock absorber was excessively damaged from venting of explosive gases through ejector tube perforations. The model crashed back to Earth. In another test, on August 5, charge canister No. 2 jammed in the ejector tube and failed to detonate; the model was damaged when it again fell to the ground. Later, on August 27, charge No. 3 was ejected successfully but failed to detonate. On August 31, for once, all charges were ejected and detonated but the shock absorber was again heavily beaten-up.

It was not until several tests later, on November 14, that a flight went off without a hitch: all the charges ejected in proper sequence and fired; the rocket rose to a height of 185 feet, and the 14-foot-wide drogue parachute that was stored in the nose of the ship was released at the peak of the trajectory and brought the model to the ground, intact. Two days later, Brian Dunne wrote an account of the flight and mailed it to Freeman Dyson, who had by that time returned to Princeton. It said in part:

> Wish you could have been with us to enjoy the Point Loma festivities last Saturday. The Hot Rod flew and flew and FLEW. We don't know how high yet. Ted, who was up on the side of the mountain, guessed about 100 meters, by eyeball triangulation. Six charges went off with unprecedented roar and precision. We think we have it all recorded.

They did: the whole sequence was caught on film, with a Fastax ultra-slow-motion camera, and the result is available on YouTube (https://youtu.be/Q8Sv5y6iHUM). Much later, after the project was partially declassified, Hot Rod was sent to the Smithsonian Institution, which put it on public display for a time at the National Air and Space Museum.

But a single successful flight of a small-scale test model offered no significant evidence that a full-size, 4,000-ton nuclear spacecraft was workable or practical. If anything, the test series showed how many different faults could emerge at virtually any stage of flight, a fact that did not encourage optimism about the prospect of a full-scale-size Orion surviving across thousands of nuclear bomb blasts without experiencing one or more of what in the rocketry business are called "anomalies," that is, breakdowns.

The lone flight success nevertheless managed to convince the Air Force to continue the project. The Orion group continued to work on it for another six years, putting forward what seemed like endless design adjustments, changes, and tweaks, and even the occasional idea for ever bigger and better vehicles.

The signing of the Partial Nuclear Test Ban Treaty in 1963 by the United States and more than a hundred other nations, effectively doomed any realistic prospect of a nuclear-bomb-powered spacecraft because the treaty banned nuclear weapons tests in the atmosphere, in outer space, and under water. Research continued until 1965, though, with the hope that the project might be exempted from the test ban, or that fission-free bombs might be developed, or that some other miracle save would emerge in the future.

In fact, speculation and proposals for new types of bomb-propelled Orion spacecraft continued to be advanced by project scientists and others even after the official end of Project Orion in 1965. In 1968, for example, Freeman Dyson wrote "Interstellar Transport," which was published in the October issue of *Physics Today*. In it he described a "conservatively designed spaceship" that would be propelled by 30 million thermonuclear bombs. Such a ship, Dyson stated, represents "the absolute lower limit of what could be done with our present resources and technology if we were forced by some catastrophe to send a Noah's Ark out of the wreckage of the solar system."

It is difficult to know how seriously Dyson himself took this idea. Indeed, some of the claims made by members of the Orion design team strain credulity beyond the tipping point. In a talk presented to the International Atomic Energy Agency in Vienna, in December 1966, Ted Taylor stated:

> I have made a morphological outline of possible space propulsion systems, classifying them according to whether the energy release is pulsed or continuous, the types of energy sources that are used, the numbers and types of energy conversion stages in the engine, and so on. If one randomly permutes the elements of this outline, one generates more than 10^{22} different space propulsion concepts, each of which makes logical sense!

This was simply beyond belief.

Herbert York, who from the beginning never thought the concept was feasible, came to a similar conclusion after the end of the program. "As a real project with real people and real money, it would have ended up with disappointed people. ... Orion involved putting together simultaneously a number of novel technologies, most of which could not be meaningfully tested in isolation from each other or on a small scale."

Brian Dunne, the chief experimentalist, reflecting on the past much later, said: "Looking back on it, I know there must have been some wild intoxication."

In the end, Freeman Dyson felt much the same: "We really were a bit insane, thinking that all these things would work." And: "I never really believed it. ... It was impossible to take the whole thing seriously, all of the time."

5   WHERE TO?

Beyond the task of developing a realistic and workable propulsion system that would make interstellar travel possible and practical, there is the prior challenge of identifying an extrasolar planet that would be suitable for long-term human habitation. Any planet that is a candidate for human colonization has to satisfy a surprisingly large number of requirements stemming from the fact that human biology has evolved on Earth and nowhere else, and is therefore fit to survive only in an environment that is substantially similar to our own. As Daniel Deudney said in his book *Dark Skies*, "Humans are sprung from the Earth, have never lived anywhere but on Earth, and the features of this planet have shaped every aspect of human life. ... Life is not *on* Earth, it is *of* Earth." And for that reason, a planet fit for human colonization elsewhere must be earthlike in several important respects.

It must be a rocky planet as opposed to one composed exclusively of liquids or gases, yet it must contain an abundance of water, an essential human need. It must be of a certain size – big enough to retain an atmosphere but not so big as to make the atmosphere overly dense and smothering. It must possess a gravitational field strong enough to hold people firmly on the surface yet not so strong that it would crush them to death. Unless potential human colonizers are to spend their lifetimes in spacesuits or inside of closed habitats, the candidate planet must have a breathable atmosphere composed of adequate levels of oxygen but without containing any toxic or other undesirable gases. The atmosphere must also filter out a significant amount of ionizing radiation that would otherwise be harmful, or even fatal, to the colonists. The planet must lie in a temperate zone throughout its entire solar year, and must orbit a star that does not send out periodic fiery super-flares that would destroy all life. For maximum comfort, it should have a livable day–night cycle. It must orbit a star that is not too dim to sustain human vision, but not so bright as to blind us.

In addition, the planet should be without extremes of geologic activity such as excess vulcanism or meteorological phenomena such as persistent and excessive cyclonic motion. It should have a balmy atmosphere – neither too hot nor too cold. It should lie in the star's "Goldilocks zone," also called the habitable zone, meaning that the planet's temperature allows for the existence of liquid water. (To say that a planet lies within the habitable zone does not imply, however, that it is habitable by humans, but merely by some form of life, even if only microbial life.)

Further, in order for people to build usable structures, the target planet must have an adequate supply of certain elements such as iron, aluminum, or other metals. It should also possess a magnetic field that helps repel its sun's solar wind from evaporating away the atmosphere. What the colonizers really need, in the words of Michael Lemonick, who wrote a book of the title, is a true "Mirror Earth," an Earth twin or duplicate. But it must not be so identically twin-like that it harbors an advanced civilization that would be likely to repulse an incoming force of human invaders that they perceive as a threat. Still, it would be difficult to rule out the existence of such a civilization absolutely beforehand, since, as Donald Goldsmith and Martin Rees argued in *The End of Astronauts*, "Advanced civilizations with no interest in being detected almost certainly possess the means to avoid detection." One does not want to arrive gleefully at an extrasolar planet only to be met by a race of super-intelligent, super-powerful, and hostile space aliens who have successfully kept themselves well-hidden for thousands or even tens of thousands of years.

In fact, there is more to worry about than confrontations with intelligent aliens. The planetary surface, or parts of it, could be covered by noxious extraterrestrial pests: by the alien equivalents of Norway rats, plus the local versions of mice, snakes, lizards, poisonous insects, fungi, bacteria, viruses, or other menacing organisms.

And as we have seen, even if, to guard against being surprised by such unpleasantness, a robotic probe had been sent out to the candidate planet prior to a human journey there, still, in the interval between (a) the robot's arrival and (b) our reception of its report, and (c) during the further, substantial interval required for the voyagers to cross the distance from here to there, conditions on the target planet might have changed to the point that would render it unfit for human life.

Since it would have to satisfy so many requirements simultaneously, an exoplanet suitable for human colonization could conceivably be very far away, making it all the more difficult, expensive, risky, and impractical for us to mount an expedition to take possession of it.

Even so, there are plenty of candidate destinations to choose from: by 2022 the NASA Exoplanet Archive had recorded more than 5,000 planets outside of our solar system. Only a tiny fraction of them, however, four percent, are "earthlike" in some sense, and many of them lie thousands of light years from Earth. One of the problems facing the selection of an appropriate destination is that of knowing, in advance, the precise physical makeup and atmosphere, if any, of the target planet. The development of spectroscopy in the nineteenth century made it possible much later, in the closing years of the twentieth century, for us to learn much about the makeup of an extrasolar planet from light years away, here on Earth. But no matter how carefully and comprehensively one has been in detecting and characterizing an exoplanet's physical features, there is always the residual possibility of being met by one or more unforeseen circumstances or conditions on the planetary surface that could not have been detected prior to our actually arriving there.

Some degree of uncertainty is inherent to any observational measurement of a planet's physical characteristics because no measuring instrument is perfect and because a substantial amount of inference is often involved in the process of characterization. Additionally, no measurement of a star or planet is ever *current* in the sense of being true at the time when the measurement is taken. When speaking of planets that are light years away from us, what we see is not what exists at the time of observation; instead, it's what the star or planet was like years ago, back when the light we are now receiving was emitted. For example, if in 2025 we observe a star that is 1,000 light years away, we are seeing what it was like in the year 1025, during the early Middle Ages on Earth, just before the Norman Conquest, not what it's like in 2025, or when we might eventually get there in 2525 or later. Our knowledge of an exoplanet's characteristics always lags behind the planet's actual, physical state in the present. Planetary conditions might have changed substantially in the interim, and changed further still if and when human colonizers ever arrive, land, and walk out onto the planetary surface. For all of these reasons, choosing the right planet for a voyage of interstellar colonization is an exceptionally risky business.

# # #

Speculation about other worlds – indeed, even about a "plurality of worlds" or "innumerable worlds" – goes back to ancient Greece, and is principally associated with the pre-Socratic philosophers Leucippus and Democritus (ca. 460–370 BCE). These thinkers (and their many followers) held that atoms are indivisible particles that exist in infinite numbers, and that all material things are made of atoms. The atoms exist in a void, and all physical objects are created by chance collisions among them. Because they are everywhere, and because these collisions happen continuously, groups of atoms mingle and merge all over the universe, and in that way create an infinite number of other worlds.

A later philosopher, Epicurus, expressed the atomist argument in this way:

> There are infinite worlds both like and unlike this world of ours. For the atoms being infinite in number, as was already proved, are borne on far out into space. For those atoms which are of such a nature that a world could be created by them or made by them, have not been used up either on one world or a limited number of worlds. ... So that there nowhere exists an obstacle to the infinite number of worlds.

This line of reasoning slightly resembles the latter-day "principle of mediocrity," which maintains, in essence, that our world is nothing special, rather that it is just average, so that there must be innumerable other worlds like our own. Such a viewpoint was expressed, for example, by astrophysicist Sebastian von Hoerner who in 1961 wrote, regarding possible extraterrestrial life in the universe: "Because we have no knowledge whatsoever about other civilizations, we have to rely completely on assumptions. The one basic assumption we want to make can be formulated in a general way: Anything seemingly unique and peculiar to us is actually one out of many and is probably average."

That assumption, however, is hardly incontrovertible. In fact, it is a fundamental error in logic to assume that the single example you have in front of you is typical or "average." It could equally well be atypical or unusual – an outlier or "unicorn." The possible uniqueness or commonness of our world is an empirical question that can be settled only by observation, that is, by the discovery or the failed detection of other worlds after having made an extensive search.

In ancient Greece at any rate the concept of an infinity of worlds was strongly opposed by Aristotle, who asserted in his tract, *De Caelo* ("On the Heavens"), that "There cannot be several worlds." This claim rested on Aristotle's belief that the universe and everything in it was composed of four elementary substances, earth, air, fire, and water, each of which had its own "natural place." The natural place of earth was at or toward the center of Earth; whereas fire, since it was light by nature, tended upward. Particles of earth tended toward the center of our Earth wherever in the universe they might be located. And for that reason those particles moved down toward the center of the Earth, which is the center of our world, so that "there cannot be several worlds."

Both the atomist and Aristotelian views concerning the problem of other worlds were equally theoretical and speculative; neither was based on empirical science, which at that time was not equal to the task of detecting extrasolar planets. But it was an accident of science history that during the whole of the period from the ancient Greeks all the way up to the middle of the twentieth century, it was still not within the power of science to resolve the question of whether exoplanets actually existed. There were two main reasons for this, the first and most obvious of which is that the stars are very far away and are optically small. Second, any planets in orbit around them are even smaller, and the light emitted by them is correspondingly dimmer. Still, the stars are nevertheless comparatively bright objects in the night sky and when the first exoplanets were discovered by astronomers it was by their observing minute optical changes in starlight.

It is an understatement to say that detecting those changes can be difficult. A case in point is that of Barnard's star and its alleged "wobble," an effect supposedly observed by astronomer Peter van de Kamp, and announced to the world in 1963.

Van de Kamp was a Dutch astronomer who lived most of his life in the United States. He held a PhD in astronomy from the University of California and yet another one from the University of Groningen in the Netherlands. In 1937 van de Kamp arrived at Swarthmore College in Pennsylvania where he became director of its Sproul Observatory. Beginning in 1938, he undertook an intensive study of Barnard's star, a rather faint red dwarf that is some six light years from Earth. He was looking for a perturbation, or "wobble," that could have been caused by a sufficiently large planet orbiting the star. The idea was that the gravitational pull of the planet would move the star slightly in one direction when at one extreme in its orbit and in the opposite direction while at the other extreme.

Such a wobble cannot be observed by gazing through a telescope in real time, but only by making a series of photographic plates taken at intervals over a period of years. Astronomers then inspect the resulting collection of plates by using a special measuring instrument that can detect changes in a star's position relative to the fixed stars around it. (This is known as the astrometric method of extrasolar planet detection.) Between the years 1938 and 1962, after more than two decades of observation, van de Kamp and his colleagues, including Sara Lee Lippincott, had taken a total of 8,260 exposures of Barnard's star.

After examining his plates van de Kamp became convinced that he had detected a side-to-side wobble of the star corresponding to the gravitational pull of an orbiting planet that was some 1.6 times the mass of Jupiter. And in 1963 he published his findings in *The Astronomical Journal* under the title, "Astrometric Study of Barnard's Star from Plates Taken with the 24-Inch Sproul Refractor." In it he concluded that, "The orbital analysis leads, therefore, to a perturbing mass of only 1.6 times the mass of Jupiter. We shall interpret this result as a companion of Barnard's star, which therefore appears to be a planet."

The announcement was major news at the time, with the media of the day running headlines such as "New Planet Discovered," "Astronomer Unwraps New Planet 500 Times Larger Than Earth," and (in the *New York Times*), "Another Solar System is Found 36 Trillion Miles from the Sun."

Astronomers at other observatories rushed to make their own attempts to confirm van de Kamp's amazing result. None could do so, however, and it gradually became clear (although never to van de Kamp himself) that this was a spurious observation. And in a feat of analysis and deduction worthy of Sherlock Holmes, a later member of Sproul Observatory, John L. Hershey, showed that the movements that van de Kamp thought he detected were actually due to small changes within the telescope itself after routine maintenance. In a 1973 paper also published in *The Astronomical Journal*, Hershey wrote that, "It is clear that the plate field of [the telescope] has experienced one or more sudden small changes in the x coordinate due to adjustment of the objective lens and change of emulsion." The "wobble" was in the machine, not in the star.

The episode was a cautionary tale that held potentially profound implications for the prospects of human interstellar migration, for van de Kamp's false positive claim about the existence of a ghost exoplanet was by no means an isolated case. There would be other such false positives in the future, as later astronomers used a variety of different techniques and instruments to detect far-flung extrasolar planets within the Galaxy. Many such claims would in the end be overturned: sometimes retracted by the claimant, sometimes questioned by others, and sometimes flatly disproven. Still other cases remain unresolved. Establishing the existence of an extrasolar planet is an iffy business, as is the task of correctly characterizing its physical nature and makeup. Interstellar travelers beware: one does not want to make the trek only to find that the planet really isn't there after all, or isn't quite like what it had been expected to be.

# # #

Indeed, the first extrasolar planets to be reported in the literature were nothing like what astronomers had expected them to be, which is to say, planets orbiting stars much like our Sun. One of the first discoveries to be announced occurred just prior to the end of the twenty-first century, in the year 1991. A group of astronomers at the Nuffield Radio Astronomy Laboratories at the University of Manchester claimed to have detected a planet that was orbiting a pulsar. This was anomalous because pulsars are rapidly rotating neutron stars, and neutron stars are super-dense objects that were remnants of a star that had exploded in a supernova and then had collapsed back in upon itself. Any planet orbiting a neutron star ought to have been destroyed by the violent explosion, and so it was puzzling how a planet could be orbiting a star that had gone supernova.

Pulsars rotate at incredibly fast rates, hundreds of times per second, and they emit bursts of radio waves, or pulses, that can be observed from Earth. In 1990, the Nuffield group, headed by Andrew Lyne, observed a Doppler shift in the arrival times of the pulses radiating from the pulsar PSR [Pulsating Source of Radio] 1829–10. The group members interpreted the Doppler shift as having resulted from a planet orbiting the star and tugging it forward and backward, toward us and away from us, alternately shortening and lengthening of the arrival times of the pulses. In 1991, the Nuffield group reported their results in *Nature*, one of the world's top peer-reviewed scientific journals, under the title, "A Planet Orbiting the Neutron Star PSR 1829–10."

And at this important juncture there came an unexpected twist to the story. As Andrew Lyne later told a journalist: "The planet disappeared." More precisely, it had never existed in the first place, as Lyne realized to his horror late one night as he was rechecking his calculations. It is common to speak of a planet orbiting a star, but this is not the literal truth of the matter: the fact is that planets and stars actually orbit around a common center of mass called the barycenter. In order to detect the observed arrival times of the neutron star's pulses, Lyne and his team first had to calculate the position of the barycenter, but as it turned out, those calculations were off.

The following January, the authors of the original paper wrote a short "Letter" of retraction to *Nature* under the title, "No Planet Orbiting PSR 1829–10." Another false positive, illustrating once again the subtle and delicate nature of the exoplanet discovery process.

The first generally accepted and true exoplanet discovery was made by the Penn State astronomer Aleksander Wolszczan and colleague Dale Frail working at the Arecibo radio telescope in Puerto Rico, in 1992. They in fact discovered not just one, but two planets orbiting the millisecond pulsar PSR1257+12, which is located at 2,300 light years from Earth. The result was published in *Nature* under the title, "A Planetary System Around the Millisecond Pulsar PSR1257+12," which revealed that the planets had masses of about 2.8 and 3.4 times the mass of Earth and moved in almost circular orbits with periods of 98.2 and 66.6 days, respectively.

But in another quirk of science history, it was not Wolszczan and Frail who received a Nobel prize for the first discovery of an exoplanet, but another team, Michael Mayor and Didier Queloz of the University of Geneva Observatory. In 1995 they discovered the planet 51 Pegasi b. Mayor and Queloz also reported their finding in *Nature*, under the title "A Jupiter-mass Companion to a Solar-type Star."

"The presence of a Jupiter-mass companion to the star 51 Pegasi is inferred from observations of periodic variations in the star's radial velocity," they said. "The companion lies only about eight million kilometres from the star, which would be well inside the orbit of Mercury in our Solar System. This object might be a gas-giant planet that has migrated to this location through orbital evolution, or from the radiative stripping of a brown dwarf."

Their detection was soon confirmed by a separate team at the Lick Observatory in California, and, later, by still other groups elsewhere. Because it was so massive a planet and orbited so close to its star, planets of this type became known as "hot Jupiters." Many more would be reported afterward, and these discoveries forced astronomers to rethink their theories of planet formation and evolution.

Why, then, was it that Michel Mayor and Didier Queloz received half of the 2019 Nobel Prize in physics "for the discovery of an exoplanet orbiting a solar-type star"? The answer is provided by Joshua N. Winn, professor of astrophysical sciences at Princeton, who asked in *Scientific American*, "Who Really Discovered the First Exoplanet?" His answer was that Wolszczan and Frail were indeed the first. "But the pulsar planets were treated as freaks, and the search for more of them has turned out to be barren and unproductive. Only one other pulsar is known to have a planet, and even in that case, the evidence is not secure."

The planets discovered by Mayor and Queloz, by contrast, orbited not a whizzing pulsar, but a "normal," solar-type star, which was a fact that astronomers were more comfortable with.

"Just as importantly," Winn added, "the discovery of 51 Peg had the same effect as the first sighting of an unexplored and seemingly boundless continent. The exponential growth of planet discoveries, and in the number of scientists working in this area, began in 1995. That's why the Nobel Committee thought Mayor and Queloz deserved the scientific spotlight (and a half million dollars)." Whether, in retrospect, that was fair to Wolszczan and Frail is another question.

# # #

Once a planet was detected, the next task for astronomers was to characterize its physical features and makeup. Some astronomers are fond of quoting the French philosopher of science Auguste Comte, who said that we would never be able to know the chemical composition of the Sun or other stars. In 1835 he wrote, regarding the stars, that:

> We understand the possibility of determining their shapes, their distances, their sizes and their movements, whereas we would never know how to study by any means their chemical composition, or their mineralogical structure, and, even more so, the nature of any organized beings that might live on their surface.

His reasoning was that the stars were so far away that there was no hope of our ever getting there and examining their chemical makeup. But in fact it is possible to do so from afar, by means of spectroscopy, the science of measuring the spectra produced by visible light or by other forms of emitted electromagnetic radiation.

Spectroscopy as a science was developed by three main figures: Isaac Newton, Joseph von Fraunhofer, and Gustav Kirchhoff. In a well-known experiment, Newton passed a ray of white light through a glass prism and found the light was refracted into a band of colors, all the colors of a rainbow: red, orange, yellow, green, blue, indigo, and violet, in that order. This happened because light waves propagate at different frequencies, and each is bent or refracted differently by the prism glass, resulting in a fanning out and separation of the white light into all the colors of the visible spectrum.

In the early 1800s, the German chemist Joseph von Fraunhofer passed light through a diffraction grating – a plate of glass containing a series of closely spaced parallel wires – and found that they, too, like a prism, produced a spectrum composed of the usual shades and, in addition, a series of dark, fixed lines. The lines were caused by specific chemical elements, and each configuration of lines and colors represented a sort of signature or "fingerprint" of that element.

Later, the German physicist Gustav Kirchhoff showed that each element produced its own characteristic pattern of lines, and he identified several of them. Incandescent sodium vapor, he found, produced a double yellow line when viewed through a spectroscope. Kirchhoff then embarked on a large project to draw a detailed map of the solar spectrum, showing hundreds of spectral line configurations, and in this way he deduced the presence of iron, magnesium, sodium, and other elements in the outer layers of the Sun. By the use of spectral analysis, astronomers have been able to characterize the chemical composition of hundreds of extrasolar planets.

Following all of the false positives and the few true positives of the late twentieth century, exoplanet discovery entered its Golden Age in the early twenty-first century with the advent of increasingly advanced instrumentation carried aboard spacecraft. The first and one of the most important of these was the launch in 2009 of the NASA Kepler Mission from Cape Canaveral. The Kepler Mission was unique in several respects. First, it would not focus only on a single star, but would instead observe more than 156,000 stars simultaneously and continuously for years at a time from a vantage point in Earth orbit high above the atmosphere. Second, it was a special-purpose mission in that its primary objective was to detect Earth-size and smaller planets that lay in the habitable zone of their respective suns. Third, the reams of data that the mission generated were beamed back to

Earth where they would be analyzed by ranks of sophisticated computers.

Across its first six months of operation, the Kepler Mission observational system discovered more than 1,200 planets, 54 of which lay within the habitable zone of the star. In addition, there were also 68 Earth-sized planets that were not in the star's habitable zone. Those that were habitable, in turn, wound up as listings in The Habitable Worlds Catalog (formerly The Habitable Exoplanets Catalog) maintained by the University of Puerto Rico at Arecibo. According to its Web page, the catalogue "is a database of potentially habitable worlds discovered by ground and space telescopes in the past decade."

The catalogue also includes a list of "the exoplanets that are more likely to have a rocky composition and support surface liquid water." These planets were ranked according to their score on an Earth Similarity Index (ESI), which was "a measure of similarity to Earth's stellar flux, and mass or radius (Earth = 1.0)." From the standpoint of possible interstellar migrations, it will be instructive to examine a few of the topmost candidates to see what kind of worlds they are, as well as exactly how much we know of them. And so we start with listing number 1: Teegarden's Star b, whose ESI was 0.95, the most Earthlike exoplanet of them all.

Teegarden's Star was named after the discovery team leader Bonnard J. Teegarden, of NASA's Goddard Space Flight Center. The names of stars and planets have somewhat diverse origins. The brightest stars that can be seen with the naked eye have proper names, such as Sirius, Vega, and Polaris, most of them having been derived from Arabic astronomers. The constellation-based names came next, as the first star maps were produced and the stars got Greek letters and then number designations, in order of brightness (magnitude). Other stars are named after the star catalogue they appear in: the planet HD 85512 b, for example, is listed in the Henry Draper catalogue, while the planet Gliese 426 b is listed in the Gliese catalogue. What a mess!

Teegarden's Star b was discovered in 2019 by a team of a frightening 150-plus scientists who published their results in the journal *Astronomy & Astrophysics*. The planet is some 12 light years from Earth, to which it was both similar and dissimilar. Similar in that it had a mass 1.05 that of Earth; it was also thought to have a rocky surface, and possibly also an ocean of water. Best of all, it was thought to have an average temperature of about 82 °F – balmy in the extreme, like Florida. It was unlike Earth, however, in that it had an orbital period of just 4.9 days, and because its sun was not a yellow star like our own, but instead was a red dwarf. Red dwarf stars are the most common type of star in the Galaxy, but are optically dim, too dim to support human vision. Most of a red dwarf's light output occurs in the infrared.

Furthermore, red dwarf stars are known to emit strong flares that could kill any possible life form on Teegarden's Star b, and could also burn away its atmosphere, assuming it had one, but its possible existence and makeup remain in dispute. The planet is also thought to be tidally locked to the star, meaning that the same hemisphere always faces the star, so that there is no day/night cycle. Thus, for all of its supposed similarity to Earth, Teegarden's Star b would not be an attractive candidate for a voyage of interstellar colonization.

Entry 2 of the Habitable Worlds Catalog is the obscure, little-known planet named TOI-700 d. "TOI" is the name of the host star and is an acronym standing for "**T**ransiting **E**xoplanet **S**urvey **S**atellite **O**bject of **I**nterest." A problem with TOI-700 d is that it is 101 light years from Earth, meaning that it would take an unacceptably long time for a band of intrepid star travelers to get there. And assuming they ever arrived, they would find a rocky planet about the size of Earth, but which is also tidally locked to its star, with one side constantly bathed in light, the other in darkness.

Crucially, however, it is not known whether the planet has an atmosphere or an ocean, and so in this case astronomers resorted to the making of models. Researchers at the Goddard Space Flight Center modeled no less than 20 different potential environments for TOI-700 d in an attempt to determine if any of them would allow for habitability. A NASA press release about their work said:

One simulation included an ocean covered TOI-700 d with a dense, carbon-dioxide-dominated atmosphere similar to what scientists expect surrounded Mars when it was young. The model atmosphere contains a deep layer of clouds on the star-facing side. Another model depicts TOI-700 d as a cloudless, all-land version of modern Earth, where winds flow away from the night side of the planet and converge on the point directly facing the star.

And these, Teegarden's Star b and TOI-700 d, are the top two candidates for habitability in The Habitable Worlds Catalog: both tidally locked to dim red dwarf stars, and neither of which is known to have a breathable atmosphere. Other members of the catalogue's top 20 candidate exoplanets lie at distances ranging from 11 to 1,193 light years from Earth. Without exception, all of the 20 orbit M-type red dwarf stars, the smallest and coolest stars, which would be unlikely to support human vision in the form of visible light. Living on any of them would be an experience of perpetually walking around in darkness or at best in artificial light.

These results considerably shrink the prospects of our colonizing any of the top 20 most habitable Earthlike stars. They may be "habitable," but not by earthlings.

# # #

Still, there is a way out of this, although it calls for some rather extreme and desperate measures. The solution is to recode the human genome so that we can see not only in visible light, but also in the infrared – and also be capable of doing other things that are not normally possible to mere humans in our current, somewhat parochial and idiosyncratic state. However unlikely all of this might sound, exactly such a reengineering of humanity has been proposed by the geneticist and computational biologist Christopher E. Mason. His notion is that we must give up the vain hope of finding a completely earthlike planet; instead we should reprogram ourselves so that we can live on one or more planets that are significantly *different* from Earth. In other words, accommodate ourselves to the planet rather than finding a planet that is a perfect fit for us.

As an expert in space science, genomics, and the reengineering of human life, Mason is, if anything, overqualified. For starters, he holds a PhD in genetics from Yale. He has been a principal investigator or co-investigator on seven NASA missions and projects, including the NASA Twins Study. He is Professor of Physiology and Biophysics at Weill Cornell Medical College, and holds affiliate appointments at the Meyer Cancer Center, the Memorial Sloan Kettering Cancer Center, and the Consortium for Space Genetics at Harvard Medical School.

In 2021, MIT Press published Mason's book, *The Next 500 Years: Engineering Life to Reach New Worlds*. The book sets out a detailed, step-by-step, radical plan for altering the human genome in such a way that would permit not us but our genetically augmented descendants to leave the Earth behind and launch toward a "second sun" elsewhere in the Galaxy within the next 400 to 500 years. More specifically, "The full, multisystem backup plan of Earth's life will initiate with the launch of the first generation ship around 2401."

Those aboard the ship, however, would be nothing like present-day humans who, Mason thinks, are unfit for living on another planet. "Sending any Earth-evolved organism to any other planet would result in almost certain death, which represents the sad, evolutionary 'good luck' plan," he says. "It is likely that we, and all other organisms, will need extensive physical and genetic help to survive anywhere else – even if just to arrive at our next destination."

For one thing, the high radiation levels encountered in deep space would require that we alter the present-day human genome to become radiation-resistant. "By leveraging currently known biological pathways for radiation resistance ... we can build newly engineered biological networks that can improve the response to radiation and the ability of humans to survive in more harsh environments."

The necessary biological pathways for radiation resistance already exist in the tardigrade, a tiny, segmented micro-animal that is known to survive in extreme environmental conditions including high heat, extreme pressure, desiccation, radiation, and in the vacuum of space. The specific tardigrade genes responsible for radiation resistance have been isolated, and Mason in his own lab is working on "embedding tardigrade genes into human cells to enable radiation resistance."

As for seeing in dim lighting, or even in total darkness, this is likewise merely a matter of engineering the human genome to allow for seeing in the infrared, giving us instant night vision: "New eyes for new planets," as he puts it.

Some other animals already have infrared vision – snakes, fish, frogs – so why couldn't we? The cones of our retinas "could be engineered to respond to additional wavelengths of light, though the interpretation of these newly absorbed photons and perception would require advanced multicellular engineering of the overall electric circuitry of the eye and brain. Solving these electrical limitations would enable an entirely new perception of the world and universe (e.g., thermographic or infrared [IR] vision)."

If "future humans will look different," as Mason says, they will also procreate differently, in "exowombs" where "the developing baby can be constantly monitored, and if/when issues arise, they can be quickly addressed and fixed." Alas, "reproduction would be further decoupled from sex."

Going farther afield, "we could even add more segments to eyes (compound eyes, as in insects), instead of being limited to just one." After that enhancement, it would not be the Martians who would be the bug-eyed monsters, it would be us. And, once we have also incorporated into our genomes the ability to photosynthesize, like plants, we'd become "chlorohumans," and our "chloroskin" cells would become green. We would not then be the little green men of science fiction, however, because photosynthesis is very inefficient at converting photons into energy and it would require a large surface area to generate enough power for a given individual. But the solution here is simplicity itself: these new chlorohumans just need to grow more skin:

> If a human epidermis was expanded 300-fold (1.7 m^2 × 300), which is about the size of two tennis courts, a chlorohuman lying on their stomach would only need to sit in the sun for about one hour. Therefore, a chlorohuman could go on a lunch break, unfurl their newfound skin in a large empty field somewhere, get a meal while maybe taking a nap, and then close up their skin and head back inside, full and satiated.

Although such a scheme may sound implausible, impossible, or even preposterous, Mason presents it straightforwardly and as if the actual implementation of these novel biological features would be easy and nonproblematic. Apparently, though, these photosynthetic chloroskin cells would need to be blood-infused, which would mean a complex system of veins and capillaries running throughout the whole double-tennis-court-size expanse of green skin. How will this be accomplished, and how will an ordinary heart be capable of fulfilling this demand? How heavy will it all be? How will it all fold up again after deployment? Where will it be carried on the body? And so on. The fact that Christopher Mason can *imagine* such stuff does not mean that this scheme is even remotely feasible in any kind of sensible or practical real-world implementation.

# # #

In assessing this brave new biologically reengineered species as envisioned by Mason, it must be said, first, that apart from whatever technological difficulties are faced by the prospect of importing these major changes to the human genome, even worse problems arise from the standpoint of their acceptance by the general culture, politicians, and by scientists themselves.

The first gene-edited humans were created in 2018 by Chinese biophysics researcher He Jiankui, who was at that time an associate professor of biology at the Southern University of Science and Technology in Shenzhen, China. He, known to his colleagues as "JK," had learned the CRISPR/Cas9 gene-editing technique as a postdoctoral fellow at Stanford. Later, JK used the CRISPR technology to insert into the genomes of two human embryos a gene that would confer resistance to HIV infection. The twin girls, known by their pseudonyms, Lulu and Nana, were born normally in October 2018.

The experiment was soon condemned by many of the world's scientists, who saw this work as irresponsible, unethical, and even a possibly dangerous use of genetic engineering technology. This despite the fact that according to He, the twins appeared to be healthy in all respects. In consequence of his having performed the experiment without having obtained prior approval from the university's institutional review board, and for having done the work in secret, JK lost his job and was prosecuted at the Shenzhen People's Court, which sentenced him to three years in prison, and fined almost $500,000. He was released from prison in April 2022.

In light of this controversy over a very slight, beneficial modification of the genomes of the two twins, it is highly unlikely that a massive human reengineering program such as proposed by Christopher Mason would be seen even by scientists themselves as at all ethically permissible. David Baltimore, a Nobel Prize-winning biologist, called He's work "irresponsible," and said that "I think there has been a failure of self-regulation by the scientific community." Other scientists foresaw potential misuses of genome editing by the military, who would be able to "weaponize" people so that they could get by on just four hours of sleep, would be pain-free despite injuries, and would have greater physical strength than is possible to normal humans.

Second, it must be observed that this ambitious project as proposed by Mason leaves us in the end with a race of creatures that are in fact no longer human. Conceived in test tubes and brought to term inside exowombs, these new chlorohumans with infrared vision, multiple insectile segmented eyes, and tennis-court-scale, photosynthetic green chloroskin – these are not humans but are instead a species of posthuman monsters. A consequence of this is that Mason's scheme fails decisively in achieving what has always been one of the primary motives for our going to the stars, which is to preserve the *human* race from extinction. Mason's grandiose project does not do that, for it is another, nonhuman species, not us, who would be going to the stars.

For us humans existing as we are today, astronomers have found no real mirror Earths, which means that even if we had the spaceship and related technologies to get to a suitable exoplanet, we have as yet no place to go. But the search continues, and one can always hope.

6 THE WORLD SHIP

Supposing that a suitable exoplanet were identified, the next issue to be confronted by proponents of interstellar travel is that of conceptually designing a spacecraft to take a select portion of the human race on a voyage of colonization to the planet in question. Advocates have proposed many different types of spacecraft to accomplish that feat, but one of the most popular, compelling, and imaginative ideas is that of the multigenerational spacecraft, also known as an interstellar ark, or world ship.

We have seen that the key component of such a craft, a habitat, but no more than that, had been described early on, by J. D. Bernal in his 1929 book, *The World, the Flesh, and the Devil*. There he invited readers to "Imagine a spherical shell ten miles or so in diameter, made of the lightest materials and mostly hollow ... The great bulk of the structure would be made out of the substance of one or more smaller asteroids, rings of Saturn or other planetary detritus."

That idea never really caught on among interstellar enthusiasts of the day, most probably because building such a ship presupposed a considerably more advanced and industrialized human occupation and use of solar system resources than was in place in the early twentieth century. Also, other than describing it in a few paragraphs, Bernal took the idea no farther: he provided no engineering plans, cost estimates, or proofs of concept. Nevertheless, this was an idea with legs because, almost 50 years later, a spherical space colony was again proposed by Princeton physicist Gerard K. O'Neill. And in 2022, *Frontiers in Astronomy and Space Science* even ran a piece about building a spherical "asteroid city" from asteroid rubble: "Habitat Bennu: Design Concepts for Spinning Habitats Constructed From Rubble Pile Near-Earth Asteroids."

None of those habitats is by itself a world ship. But the kernel notion underlying the world ship concept is nevertheless that of a space colony; indeed, a world ship is essentially nothing more than a self-sufficient space colony moving fast and headed toward an attractive and inhabitable planet elsewhere. And the figure most closely associated with space colonies is Gerard K. O'Neill, whose 1977 book, *The High Frontier: Human Colonies in Space*, brought the concept to wide public attention. His earlier, 1974 paper, "The Colonization of Space," published in *Physics Today*, was influential among scientists, and in 1976 physicist Gregory L. Matloff went a step farther and published an article in which he explored the notion of fitting an advanced propulsion system to an O'Neill-type colony, thereby turning it into an interstellar ark. The plausibility of constructing a world ship thus hinges on the prospects of creating an O'Neill-type space habitat, or something substantially similar, to which a propulsion system is added.

Gerard K. O'Neill was an important American physicist who made fundamental contributions to many areas of science and technology. While a professor at Princeton University he invented the particle storage ring, a system for confining high-energy particles before releasing them into the main ring of a particle accelerator where they would be collided with other particles at higher energies than previous accelerators attained. He also played a role in the creation of the spaceborne global positioning system.

O'Neill was an accomplished pilot of both powered aircraft and sailplanes, and had been interested in space and space travel since he was a boy. In 1966, when NASA established a new category of scientist-astronaut, O'Neill applied for acceptance into the space program. He traveled to San Antonio, Texas, to the School of Aerospace Medicine at Brooks Air Force Base, for a week's worth of physical and mental tests. In all, he thought he had a pretty good chance of making the cut. And then in 1967, when he got a call from Alan Shepard, one of the original Mercury Seven astronauts and the first American in space, O'Neill was almost sure he'd made it. But when Shepard got to the end of his spiel, O'Neill heard the words "unfortunately," and "I'm sorry to tell you," and realized all at once that this was the end of his budding career as an astronaut.

Two years later, in 1969, Neil Armstrong and Buzz Aldrin walked on the Moon, and O'Neill got newly excited about space and space travel. That fall, while teaching a seminar at Princeton, he put a question to his students: "Is a planetary surface the right place for an expanding technological civilization?"

That was a fairly odd question: it was akin to asking: "Is skin the right covering for the human body? Why not Kevlar, or Gore-Tex, or Nomex?" As it was, O'Neill himself was never really certain as to why he asked that specific question, which he initially regarded "almost as a joke."

"There is no clear answer," he said, "except to say that my own interest in space as a field of human activity went back to my own childhood, and that I have always felt strongly a desire to be free of boundaries and regimentation."

He and his students took the question seriously, however, and together they started researching what it would take to establish human colonies in Earth-orbital space. The answer was: a lot. "To live normally," O'Neill later wrote, "people need energy, air, water, land and gravity. In space, solar energy is dependable and convenient to use; the Moon and asteroid belt can supply the needed materials, and rotational acceleration can substitute for Earth's gravity."

O'Neill's students went on to make calculations, write papers, and theorize about how best to do all of this, as did O'Neill himself. And at the end of it, they had collectively decided that an immense, luxurious, self-sufficient, and all-inclusive colony could be built in orbit. Furthermore, it was all doable with existing, or near-future technology, it wouldn't take that long to build, and it wouldn't cost all that much. Or so they thought.

They ended up describing two different geometries: a sphere, much like a Bernal sphere, only smaller, and a pair of counterrotating cylinders. The cylinder configuration is the structure most closely associated with the idea of O'Neill colonies, and with good reason. It is the only configuration mentioned in "The colonization of space," where, under the heading "A cylindrical habitat," O'Neill says: "The geometry of each space community is fairly closely defined if all of the following conditions are required: normal gravity, normal day and night cycle, natural sunlight, an earthlike appearance, efficient use of solar power and materials. The most effective geometry satisfying all of these conditions appears to be a pair of cylinders." He puts the size of each cylinder as "about four miles in diameter, and perhaps about sixteen miles in length." The total population of the two cylinders together would be 10,000 people.

Notwithstanding the importance of the O'Neill cylinder, there is no doubt that in *The High Frontier* O'Neill describes the first space colony, Island One (called Model 1 in "The Colonization of Space"), as a sphere with a circumference of nearly a mile as satisfying all the requirements of a human habitat in space: "For a given volume enclosed, a sphere is the shape that requires the least surface area. That is important for minimizing the required mass of cosmic ray shielding," O'Neill wrote. "If the sphere rotates at 1.97 rotations per minute, it will provide Earth-normal gravity at its equator, near which most apartment areas can be located."

Similarly, *The High Frontier* also describes a second, larger colony, Island Two, as a sphere, having "an 1,800-meter diameter (about 6,000 feet) with an equatorial circumference of nearly four miles. Island Two, with about ten square kilometers of total internal surface area, could house and maintain a population of 140,000 people, possibly in a number of small villages separated by park or forest areas."

Island Three, the largest of all those O'Neill proposed, is the only colony in his book that he describes as cylindrical. "'Island Three' could have a diameter of four miles, a length of twenty miles, and a total land area of five hundred square miles, supporting a population of several million people." Given that much land area, such a population is not unrealistic: the borough of Manhattan houses a population of more than a million, although its land area is less than 23 square miles. (This curious duality of O'Neill colony architectures – sphere vs. cylinder – has been little noticed in the subsequent literature.)

O'Neill routinely characterizes all of these colonies, whatever their shape, in the most glowing terms. The living spaces comprising the interiors would be nothing like the sterile, hard-surface, stainless-steel enclosures such as were found inside the Space Shuttle and the International Space Station. "Each family of five people can enjoy a private apartment of about 2,500 square feet area with a private, sunlit garden of some six hundred square feet," O'Neill writes. "By arranging the apartments in terrace fashion, most of the remainder [would be] available for parks, shops, small groves of trees, streams, and other areas available to all inhabitants." These inhabitants were not limited to people, but also included animals and birds. Indeed, it would be "an environment like the most attractive parts of Earth."

Sunlight would shine into the colony through strips of windows that alternated with the land areas verdant with lakes, rivers, trees, and grass – everything but waterfalls. The inhabitants would enjoy many of the activities ordinarily pursued on Earth, and also some that aren't. "All Earth sports, as well as new ones, are possible in the communities," O'Neill says. "Skiing, sailing, mountain climbing (with gravity decreasing linearly as the altitude decreases)."

The colony's interior atmosphere would be that of sea-level air pressure on Earth, and the residents would be protected from cosmic rays by the thickness of the outer structure. Bountiful amounts of energy would be available from the Sun.

Most of the raw materials needed for construction would be mined, refined, and fabricated on the Moon and shot up to the colony by an electromagnetic launch system. This would be true even of the atmosphere: "In order to minimize costs, probably the early inhabitants will have atmospheres composed of the material most plentiful on the Moon: oxygen," O'Neill writes.

Further, "I have calculated that the first model community would require a construction effort of 42 tons per man-year, comparable to the effort for large-scale bridge building on Earth." The construction would take place in space, and the colonies would be located at the Lagrange point L$_5$, where the mutual gravitational fields of the Earth and the Moon are in balance and cancel each other so that the colony would float free without any need for rocket thrusters or active station-keeping.

The total mass required to build the Model 1 space colony O'Neill estimates as more than 500,000 metric tons. Despite this enormous bulk, in his *Physics Today* piece, O'Neill gave an "earliest estimated date" for completion of Model 1 as 1988, just 14 years after its 1974 publication date.

As to the cost of it all, O'Neill writes that, "The initial estimates have been aimed at holding the cost equal to that of one project we have already carried through: Apollo," which he put at $33 billion. Just three years later, however, in *The High Frontier*, the project had already become much more expensive: "The general estimate is from two to five times the cost of the Apollo Project," or about $60 to $150 billion.

# # #

After he'd first written "The colonization of space," O'Neill had had some trouble in getting the piece published. He sent it to *Science*, whose reviewers said no. He sent it to *Scientific American* with the same result. In fact, he spent a total of four years successively submitting his piece from one journal to the next, getting rejection after rejection. This is quite understandable: it was an article, after all, that described a pair of cylinders, each four miles in diameter and 16 miles long, floating in space and jointly housing some 10,000 human residents. For many readers, that was a lot to accept at face value. Finally, Harold Davis, the editor of *Physics Today*, said yes, but even he asked for some clarifications and revisions. Meanwhile, O'Neill gave lectures at colleges and found that students were absolutely gung-ho about the idea of colonizing space. At length the whole concept became so widespread and popular that he wound up testifying before subcommittees of both the US House and Senate.

O'Neill's Model 1 got conceptually converted into a world ship in 1976, with the publication of Gregory Matloff's paper "Utilization of O'Neill's Model 1 Lagrange Point Colony as an Interstellar Ark" in the *Journal of the British Interplanetary Society*. Matloff took the idea quite literally, as he envisioned coupling the space colony, essentially unaltered, even down to its windows ("solars'), to a Daedalus-type, nuclear fusion propulsion system placed between the two cylinders. (In *The High Frontier*, O'Neill writes that to think of taking his colonies to the stars "we must carry our speculations well beyond the limits of present-day technology," suggesting, for example, an antimatter-based propulsion system.)

But before taking O'Neill colonies to the stars, it is first necessary to assess the feasibility of the colony itself, physically, financially, and politically. The colony as a distinct physical structure does not appear to violate any known laws of nature and so it must be said to be possible in principle. But what is theoretically possible is not the same as what is politically possible, financially feasible, practical, or what makes good sense. In the case of O'Neill-style space colonies it is questionable whether their cost would be worth whatever benefits were to be realized by them.

First, the experience of living one's life inside a closed sphere or cylinder may be so much at odds with everyday human experience, as well as expectations set by human nature, that it would not receive wide acceptance among ordinary earthlings. O'Neill's assumption is that the cylinders will be tolerable to residents who, when they look up, will not see the blue sky above, but rather a bunch of houses, lakes, rivers, and forests hanging upside down over their heads. But in fact that upside-down, outside-in vista might be more likely to induce disorientation, claustrophobia, or nausea than to inspire the kind of delight verging on euphoria that O'Neill himself routinely exudes and attributes to others. Further, the rotation of the colony might produce vestibular effects and "space sea-sickness" among some residents.

Whether the colony would be feasible politically would depend heavily on its cost, and it is here that O'Neill is most vulnerable to criticism. His initial paper included a chart of "Masses of Materials Required for Model 1," expressed in units of metric tons. Although he regularly spoke of Model 1's total population of 10,000 in both cylinders combined, the chart lists only "2000 people and equipment," with a total mass of 200 metric tons, as being lifted from Earth to Lagrange point L_5, where the construction would take place. But those 2,000 people are only *the workers who build the structure*. Thus, in O'Neill's cost analysis there was an undercount of 8,000 people who would also need to be transported upward to the completed Model 1, and that cost cannot be dodged or dismissed.

Second, there is O'Neill's figure for the cost of lifting a pound of mass to L_5, which he gives as $425 per pound, without saying how he arrived at that figure. He further claims that, "Around the year 2000, a fully reusable chemical rocket system could transport payloads to L_5 at a cost of about $100 per pound," and adds that, "The near certainty of continued advances in propulsion systems suggests that the actual costs will be lower."

But just the opposite actually occurred. In 2021, a NASA Advanced Space Transportation Program fact sheet stated that, "Today, it costs $10,000 to put a pound of payload in Earth orbit." In 2022, Elon Musk's SpaceX, which, unlike NASA, does use reusable rockets, was charging $1,200 per pound to fly to low Earth orbit. In either case, O'Neill's figures and drastic lower cost predictions are wildly overoptimistic.

The dollar figure given by O'Neil in his initial 1974 paper for building the Model 1 space colony, $33 billion, covers only the materials and people to be lifted *from Earth* to L_5. O'Neill assumes that the bulk of the materials required for construction of the colony will come from the Moon: "98% can be obtained from the Moon," he says. These include aluminum (20,000 metric tons), glass for the solar panels (10,000 mt), and water (50,000 mt). And all of this additionally assumes the existence on the lunar surface of a sophisticated mining, smelting, and fabricating operation, plus one or more mass drivers to propel the finished materials to L$_5$, plus, furthermore, a work force necessary to run these facilities, plus the housing, food, and all the rest of the needs and paraphernalia of a small working community resident on the Moon.

In 2008, *The Space Review* ran "Revisiting Island One," a piece that assessed O'Neill's cost estimate for building the Model 1 space colony, and said:

> O'Neill's expectations about launch costs ... proved to be highly optimistic, even given the disagreement about how these are to be calculated. A $10,000 a pound ($22,000 per kilogram) Earth-to-LEO [low-Earth-orbit] price, almost twenty-five times the estimate O'Neill worked with, is considered the reasonable optimum now.
>
> Going by those numbers, and the relatively greater difficulty of routine trips out to L-5 and the lunar surface that O'Neill suggests, the price of getting Island One going would be a staggering $1.5–10 trillion.

It is important to realize that these numbers, although they here pertain only to methods and designs proposed by O'Neill, would probably pertain, with certain modifications, to the construction of other world ship design habitats (not including their propulsion systems) provided that they were of relatively comparable size and building materials, and contained a population of similar size to that of Model 1.

O'Neill's proposals were in vogue, among scientists, politicians, and the general public for a time, but gradually fell out of favor and are infrequently discussed today. Part of the change of attitude was due to the work of US Senator William Proxmire. Proxmire was chairman of the senate subcommittee responsible for NASA's budget. He was also the creator of the "Golden Fleece Award," which, starting in 1975, he conferred on projects that he saw as wasteful of taxpayer dollars. For example, he bestowed a Golden Fleece Award upon the National Science Foundation for their having appropriated $84,000 for a study of why people fall in love. Proxmire presented another such award to the Justice Department for conducting a study of why people wanted to get out of jail.

In 1977, after the CBS program *60 Minutes* featured a segment about O'Neill's space colonies, Proxmire issued a statement saying: "It's the best argument yet for chopping NASA's funding to the bone. As Chairman of the Senate Subcommittee responsible for NASA's appropriations, I say not a penny for this nutty fantasy." Later, when it became clear that government support for such a project was politically untenable, public enthusiasm for O'Neill's colonies began to wane and wither away, although the general concept of a space colony itself has endured through time and has become an inescapable fixture of our overall conceptual environment.

# # #

As large as O'Neill's "nutty fantasies" were to begin with, subsequent designs for world ships only got bigger. Much bigger.

In 1984, the *Journal of the British Interplanetary Society* published a special issue devoted to "Project World Ship." Two of the papers in this issue were remarkable for the magnitude, scale, and general unrealism of the spacecraft that the authors proposed. It was as if the authors accepted no limits upon their imagination, or on the concept of limits of any kind. Indeed, most of what they said in their papers was astonishing beyond belief.

The articles in question were A. Bond and A. R. Martin, "World Ships – An Assessment of the Engineering Feasibility," and A. R. Martin, "World Ships – Concept, Cause, Cost, Construction and Colonisation." The authors of the new designs were BIS members Anthony Martin and Alan Bond, both of whom at the time of writing were engineers working on nuclear fusion technology at the Culham Laboratory in Abingdon, England. They carefully disassociated their views and proposals from their work at the lab, saying in a footnote: "The authors would like to note that this work is a private venture, and is in no way connected with their duties at Culham Laboratory."

The first study, concerned with the engineering feasibility of the craft described, started off with a number of disclaimers, provisos, and qualifications:

> The main problem is coming to terms with the scale of the concept. It is relatively straightforward to calculate the structural and propulsive requirements of a vehicle with dimensions on the order of hundreds of kilometers, but one must then face the question of the level of manufacturing effort required to build it.
>
> Today, no industrialized nation, or group of nations, could even begin to muster the necessary production effort. However, it is well within the limits of projection that some centuries hence, aided by nuclear and solar power, machine intelligence, and the abundant resources of the solar system, such a capability will become available.
>
> From the outset, then, it must be realized that the following discussion is not concerned with events which could occur within the next few centuries. But equally important, the first world ships could be constructed within less than the next millennium.

It soon became clear what the reason was for all of these caveats, for the authors went on to describe a world ship that would carry a total population of 250,000 individuals. And that was only the bare minimum. In the second paper, by Martin alone, the author states that "the carrying capacity of the largest vehicles proposed is only some 50 million (that is, equivalent to the population of the British Isles)."

The first paper considered various design configurations of the "smaller" spacecraft, including a sphere, a cylinder, a torus, and a cylinder of stacked toruses. Bond and Martin stated that the spacecraft must model the Earth as closely as possible and must be precisely suited to *Homo sapiens* as a species. For that reason, all configurations would provide for artificial gravity by rotating the vehicle about an axis of symmetry. In the end, they opted for a simple cylindrical world with a diameter of some 14 to 20 kilometers and a length of 90 to 120 kilometers. In fact they proposed *two* such ships, a "dry" world ship, composed mostly of land area, atmosphere, and passengers, and a "wet" ship of about the same size, carrying water, to provide for an ocean on the new planet. This "oceanarium" would allow for "the accommodation of a wide range of ocean dwelling species." To shield against radiation, the dry ship would have a hull some three to six meters in thickness, and would be made of steel derived from the asteroid belt, plus a layer of lunar regolith.

What is more, an actual mission to the stars would not be undertaken by merely a single vehicle traveling alone, but would consist rather of a flotilla or fleet of such ships. To preserve maximum fidelity to Earth, each craft would model a different specific earthly environment, for Earth has not just one type of environment but many.

The propulsion system according to the designers would consist of multiple pulse units of frozen hydrogen, all with solid deuterium cores. Each such unit would be ejected from the world ship and would be detonated, somewhat after the manner of the Project Orion spacecraft, but at a distance of some miles away from the habitat module. The propellant mass alone would total 778 billion tons. The craft would travel through space at 0.5 percent the speed of light, which would allow travel to the nearest stars within a thousand years.

After having provided further design details, equations, and calculations, the authors ended by saying: "It is concluded that very large vehicles are practical, on the evidence presented, although several centuries of economic growth, and a Solar System based economy, will be required before construction of such vehicles could be commenced. ... In summary, therefore, it has been shown that there are no overriding grounds for rejecting the concept of world ships."

The second paper, written by Anthony Martin alone, addressed, among other things, the economics and cost of world ships. He furnished an estimate of $10,000,000,000,000 (10 trillion USD), "in very round figures." Martin gave no accounting of how he arrived at that number, which therefore must be taken as little more than a stab in the dark. He also predicted that, "At two percent growth per year, the first interstellar probes would be built in about 230 years time. If world ships are 1,000 times more difficult an undertaking [i.e., than the first probes], then it would require 580 years for their realization. . . . And the first mass population emigrations from the Solar System may take place around 2500–3000 AD."

Elsewhere in his paper, Martin mentioned a higher population per world ship than he and Alan Bond had used previously, now saying that "World ships may contain 400,000 to 700,000 people," evidently pulling these numbers out of a hat.

It is difficult to know what to make of all this, other than to say that these are some of the most alarming and unrealistic thoughts ever conceived and expressed in words by serious scientists.

But as it turned out, other researchers later went on to offer even larger and yet more implausible designs for interstellar vehicles, almost as if this was a battle of wits to propose the most grotesquely oversize world ship in history. And so in 1996, writing in the journal *Nanotechnology*, one Thomas L McKendree discussed what would be possible if materials provided by molecular nanotechnology were used to build spacecraft in place of then current structural building materials such as aluminum, steel, and titanium. Molecular nanotechnology was the theoretical ability to design and build products to atomic precision. Such a technology, which does not exist as yet and might never, would allow the use of diamondoid materials that had much higher strength-to-density ratios than those that are now used to build structures. In his paper "Implications of Molecular Nanotechnology Technical Performance Parameters on Previously Defined Space System Architectures," McKendree argued that the use of diamondoid structural materials would make possible *extremely* large space colonies. The classic cylindrical colony, for example, if made of diamondoid structural elements could have a radius of 461 kilometers and a length of 4,610 kilometers, or 2,865 miles. That is about the driving distance from Oregon to New Jersey. According to McKendree, the habitable area of a cylinder of that size "yields a possible population for this structure of 99 billion people."

If Bond and Martin's spacecraft were alarming, McKendree's was berserk at some deep, undefined primal scream level.

#

A key issue concerning the design of any interstellar ark is the minimum population size necessary for the crew to be able to survive the journey and to establish a new civilization elsewhere. That question has been addressed by several researchers whose answers, unfortunately, vary by exceptionally wide margins.

One issue *not* much addressed by interstellar travel advocates, however, is whether human procreation in a deep space setting is a viable undertaking across hundreds of years of travel time. Certainly no human conception is known to have occurred in space, much less a human birth. A 2023 article in the journal *NPJ Microgravity*, "Human Development and Reproduction in Space – A European Perspective," claims that, "Currently, there is limited knowledge available on the systemic effects of spaceflight stressors, e.g. altered gravity (micro-, hypo- and hypergravity), increased radiation, social isolation, confinement, sleep disturbances, dietary changes and any associated stress/anxiety on the hypothalamic-pituitary-gonadal (HPG) axis in females and males and how these stressors impact the functionality of the reproductive organs." These factors constitute a very substantial known unknown, but in the face of this fact interstellar advocates simply take it for granted that human procreation in space is possible and nonproblematic, and then go on to make minimum population estimates. But in fact, this particular unknown constitutes a potential barrier to the entire enterprise of multigenerational interstellar travel.

Notwithstanding this considerable obstacle to the concept of multigenerational travel through space, in 2002 the American Association for the Advancement of Science (AAAS) held, in Boston, a symposium on interstellar travel, specifically focused on multigenerational vehicles. University of Florida anthropologist John H. Moore produced a study called "Kin-Based Crews for Interstellar Multi-Generation Space Travel." The author first summarized the diversity of scenarios presented by interstellar travel advocates: "Space crews in various conditions of animation, representing various combinations of ages, sexes and genders, often accompanied by assemblages of frozen embryos, egg cells, sperm cells and body parts, and sometimes robots and cyborgs, somehow survive and reproduce over several generations and hundreds of years to reach a habitable destination."

The author found some of the proposed trip scenarios bizarre, as they required social structures and an intensity of human relationships that were beyond what is regarded as normal and acceptable in human affairs. Moore himself, by contrast, wanted to define a more realistic mobile community that would be stable and sustainable for 200 years, or six to eight human generations. For reasons of economy, the spacefaring community should be as small as possible, but at the same time it should be one that exhibits as much genetic variability as possible. He suggested using a conventional nuclear family as the basic structural unit.

Moore further outlined a set of guidelines – in effect a social compact – for an optimal marital regime aboard the ship. These were, first, that every crew member would have an opportunity to marry, and be able to choose among at least 10 possible spouses, none of whom would be more closely related to each other than second cousins. Second, everyone would be permitted to have children.

The question was, how many individuals would make up such a starting population whose size would be stable across time? To answer it, Moore had recourse to "a population program called ETHNOPOP©, just developed by my research group." The computer program used various combinations of age and sex distributions, death rates and birth rates, in order to run a "population performance simulation." Testing populations of different initial sizes, the program found that a population of 150 to 200 would sustain itself at that size indefinitely. Still, when Moore ran the program further to simulate hundreds more years of travel, he found that a population even half of that size, or about 80 people, would suffice equally well. This result, Moore said, agreed with "the experience of small-scale societies on Earth," such as Polynesian seafaring colonists who set out to find unoccupied Pacific islands elsewhere. So the result of his analysis was that an ideal interstellar population size ranged from 80 to 200 individuals.

Later, in 2014, another anthropologist, Cameron Smith, published in *Acta Astronautica* an "Estimation of a Genetically Viable Population for Multigenerational Voyaging." He surveyed the literature on minimal viable population sizes for various animal species and found that "previously proposed such populations, on the order of a few hundred individuals, are significantly too low to consider based on current understanding of vertebrate (including human) genetics and population dynamics."

And then, taking into account several issues in population genetics such as DNA mutation rates, genetic drift, and other relevant factors, he too used computer modeling to determine the size of a population that could survive in good health aboard a multigeneration voyage lasting 150 years. Smith came up with much higher estimates than Moore did, ranging "anywhere from roughly 14,000 to 44,000 people." That range in itself represented quite a large degree of uncertainty.

More recently, in 2018 astrophysicist Frédéric Marin and particle physicist Camille Beluffi, writing in *JBIS*, published "Numerical Constraints on the Size of Generation Ships," in which they used a Monte-Carlo-based computer simulation to determine the size of an ideal starting population. They came up with two numbers: 150 individuals and 98, saying, "we emphasize that the minimum crew of 98 settlers we found is a lower limit," and acknowledged that further work might well suggest a larger figure.

Finally, in 2020, writing in *Scientific Reports*, Jean-Marc Salotti produced a study on the "Minimum Number of Settlers for Survival on Another Planet." Using computer modeling "based on the comparison between the time requirements to implement all kinds of human activities for long term survival and the available time of the settlers," the author stated that, "The minimum number of settlers has been calculated and the result is 110 individuals."

In the wider, quite extensive literature on the subject still other researchers from various disciplines, but all of them using special-purpose computer modeling, generated yet additional population estimates between the lower limit of 80 and the highest estimate of 44,000 people. In the entire published literature, no two authors came up with the exact same number, although there was a preponderance of smaller size population estimates of 80, 98, 100, and 110 – which are in fairly good agreement. In all this, however, there is a striking discrepancy and an extreme disconnect between the size estimates for a world ship *population*, and the size of a world ship *carrying capacity*, which as we have seen ranges from 100,000 to 250,000, to "millions," and even billions.

What to make of this wild mishmash of competing interstellar ship sizes and designs, and the widely diverging estimates of their ideal starting populations?

# # #

In 2020, interstellar researchers Andreas Hein, Cameron Smith, Frédéric Marin, and Kai Staats, undertook what is arguably the most comprehensive, authoritative, and impressive review and assessment of multigenerational starship concepts produced to date. This was "World Ships: Feasibility and Rationale," published in the journal *Acta Futura*. The authors discussed the definition of "world ship," addressed the central problem of differing population sizes as advanced by multiple researchers, evaluated the economics of an interstellar world ship voyage, and compared world ships to other types of interstellar vehicles.

Drawing on an extensive literature review of more than 70 scholarly papers on the subject, the authors defined a world ship as one that would be self-sufficient for thousands of years, holds a population of greater than 100,000 individuals, and travels at less than one percent the speed of light.

The authors laid down "an important precondition" underlying the entire enterprise, to wit: "we must assume that habitats in which humans can live out multiple generations can be constructed." But that is a lot merely to *assume* at the outset, as this is one of the very points at issue. It is something that must be proven, or at least established as likely, not just assumed, presupposed, or accepted without argument.

But once thus assumed, the authors go on to qualify and hedge some of their claims, saying, for instance, that, "The population size should be taken as order of magnitude values and are somewhat arbitrary." (Agreed.) They noted further that the criteria used to establish population sizes are themselves somewhat "fuzzy," and emphasized that by making a given ship design bigger, "a population of 100,000 to one million can be accommodated without fundamentally changing the nature of the spacecraft."

The authors also introduce a multigenerational vehicle design not previously discussed, the Enzmann starship, as conceived by Robert Enzmann in 1964. This spacecraft is unique in that "the population size does not stay constant but increases 10 times during the trip." It would consist of 200 people at the start but would increase to 2,000 inhabitants during the course of the voyage.

As to the matter of conflicting world ship sizes and designs, the authors said that, "These differences are a result of different assumptions regarding the size of the habitat." More particularly, they reflect the underlying suppositions on the part of the respective designers rather than being founded on anything like a rigorous scientific principles, proofs, or arguments. "It is natural that a variety of population sizes have been proposed for D1 [deme 1], the Earth-departing founding population, as researchers from different backgrounds have brought various approaches to this question."

It was as if these variations and approaches were essentially matters of differing philosophies, tastes, or styles – in other words sheer preferences – as opposed to conclusions based on objective evidence or argument.

The most important and telling point made by the authors, however, concerned the issue of reliability of the ship and its component parts. "World ship reliability is likely to be a major feasibility issue, due to the large number of parts and the long mission duration." And, we might add, the great complexity and interdependence of those components.

"Key roadblocks for world ships are the large amount of required resources, shifting [their] economic feasibility beyond the year 2300 and the development of a maintenance system capable of detecting, replacing, and repairing several [failed] components per second."

It is this last requirement – the ability to repair or replace failed mechanisms quickly and repeatedly across hundreds or thousands of years of operation – that most seriously calls into question the viability of multigenerational interstellar vehicles.

In an earlier (2012) analysis published in *JBIS*, "World Ships: Architecture and Feasibility Revisited," Andreas Hein together with still other coauthors, gave an estimate of how many parts would fail per unit of time on the Bond and Martin world ship previously described. Using data pertaining to the mean time between failure rates for aircraft and spacecraft parts, and scaling up the numbers to the replaceable parts count of the world ship, the authors calculated that to achieve a reliability value of 99.99 percent, an average of three parts per second would have to be replaced. "After 50 hours," they said, "about 567 thousand parts have to be replaced." (These numbers do not add up. Three parts per second equals $3 \times 60 = 180$ per minute, which means $180 \times 60 = 10,800$ parts per hour, which in turn translates to $10,800 \times 50 = 540,000$ parts per 50 hours.)

By any standard, that is a staggering number of parts to replace on a regular, ongoing basis. One could of course reduce that number by accepting a lower reliability value, but that in turn would also reduce the probability of the voyage being successful. There is no way that the human crew could keep up with the necessary replacements in order to achieve 99.99 percent reliability. Therefore, an automated system must detect failed parts and replace or repair them.

"The biggest feasibility issue is probably to sustain a large and complex repair facility which is largely autonomous. Furthermore, this repair facility will be subject to repair itself and thus has to be self-repairing."

And thus by this means we arrive at the major stumbling block facing the entire multigenerational enterprise, namely the fact that any automated self-repairing system is *itself* an electromechanical apparatus that is susceptible to the same type of component failure or malfunction as the components it was designed to repair or replace. Given that any physical apparatus, no matter how robust, fault-tolerant, and resilient, can fail, the automated repair system itself is highly likely to fail sooner or later, as it is arguably one of the most consistently operational physical systems aboard the ship. And given its intended purpose – which is to keep the ship running and functional – it is perhaps the most crucial and indispensable such system. And if that is true, it means that the failure of the ship's self-repair capability effectively dooms the ship, the voyage, and its passengers. The possibility of such failure, therefore, calls into question the soundness of the idea that a world ship can operate successfully across very long stretches of space and time.

The only way out of this self-referential conundrum would be for members of the crew to repair the failed self-repairing mechanism. But that assumes that those crew members have the knowledge, skill, and parts to put the system back together again, and to do so at a fast enough rate to stanch the flow of other failures that continue to occur at a rapid, three-per-second, 10,800 parts per hour rate while the automated self-repair system is inoperable. Such a replacement rate by crew members is almost certainly unsustainable.

It is worthy of note that precisely this type of potentially catastrophic occurrence has been depicted in the context of a "hard science fiction" setting. Hard science fiction is a narrative form that adheres to the known rules and laws of science. In his 2015 novel, *Aurora*, hard SF author Kim Stanley Robinson describes a multigenerational voyage to another solar system. The ship in question makes repairs by means of automated 3D printers, and those printers are themselves self-repairable since the printers can print out additional printers.

"You can print DNA and make bacteria," the narrative says. "You can print another printer. You could print out all the parts for a little spaceship and fly away if you wanted."

So far, so good. "Then one day one of the printers breaks." Then another one fails, and then another. . . . "Or maybe it's all the printers at once. They aren't working."

The problem is that all of the printers are controlled by a quantum computer, and "it seems a gamma ray shot through the ship and made an unlucky hit, collapsing the wave function."

A crew member thus has to repair the quantum computer, but a further problem immediately arises: "There's no one in this ship who really understands it." And therefore there's no one who can repair it. So unless some deus ex machina miracle entity or force magically repairs the quantum computer and saves the day, this particular voyage to the stars is doomed. And so too it would be in real life: if and when the same or similar circumstance befalls the ship's self-repair system, the voyage will have the same dire outcome.

It must be said, however, that the breakdown of the self-repairing system is not inevitable, although it is extremely likely to occur. Arguably, at least some shipboard systems and components will escape failure across the entire trip, and the automated self-repairing system just possibly could be one of them, however improbable that would be. Anyone who owns a car, washing machine, or dishwasher knows that, sooner or later, things fail. Thus, the automated self-repairing system remains as one of the ship's greatest assets and also its greatest weakness.

This, then, is the situation. The whole business of multigenerational travel presupposes that human reproduction can occur across the ages nonproblematically, for which there is no evidence, and which there is no good reason to believe. What we have in the concept of a multigenerational interstellar vehicle, furthermore, is a conglomeration of several competing designs, ship sizes, and "somewhat arbitrary" starting population estimates. Some of the larger ships are so outsize that they require a materials resource base and an economic foundation that embraces the entire solar system. And one of the proposed ships, the McKendree cylinder, is capable of holding a population of 99 billion people, who do not even exist.

While all of this, or at least some of this, might yet be possible in principle, a multigenerational world ship in flight is a risky proposition at best, and is far more likely to mechanically founder, self-destruct, and exterminate the crew, than to reach its destination complete and whole with all aboard alive and well.

# # #

Nevertheless, let us suppose for the sake of argument that the unlikely happens, and that after traveling for hundreds or thousands of years and seeing many generations of descendants rise up, live out their lives, and then perish aboard the ship, the remaining extant travelers arrive at their new home, only to discover, with considerable shock and amazement . . . an entire human civilization already living and thriving there, with a rich and fully developed culture! Members of this extrasolar human tribe now welcome them with open arms!

"We knew you were coming!" they say. "We've read all about you in our history books."

And then there is the bad news: "Unfortunately, some or all of the crew might harbor transmissible pathogens that we'd probably have no immunity to, and so the whole lot of you must be quarantined for ninety days."

But how could this possibly happen? No, the humans who were already there did not spontaneously materialize and evolve on the new planet. Given the random, opportunistic, and branching nature of evolution by natural selection, the birth of a parallel and newly evolved human species on another planet is an almost impossible scenario.

The explanation is rather that during the hundreds of years that the passengers were traveling through space, humans back on Earth had invented a series of much faster interstellar spacecraft than the multigenerational monstrosity now arriving. The fastest of the new spacecraft had long since departed from Earth orbit and had overtaken and surpassed these miserable multi-generational slow pokes and had beat them to the new planet, making their entire journey redundant and unnecessary, if not completely pointless. They needn't have gone, needn't have wasted all those multiple generations inside what was tanta-mount to a prison. They should have waited back on Earth, waited until those faster ships were developed.

And thus we come to the problem known in interstellar theorizing circles as "the wait calculation." When, if ever, is there an optimal time to launch a voyage to the stars? Leave too early and technological progress might give rise to a faster ship that will overtake the early starters. But waiting too long might mean that a human presence on another planet would occur later than it otherwise could have.

But since the pace of future progress in spacecraft propulsion technology is unknown, one cannot know in advance how long to wait. There is never a strong incentive to leave ... except if human civilization on Earth were in clear and present danger of destruction. In that case, however, available resources might already be stretched so thin in the attempt to cope with the existential threat that it would be difficult or impossible to build a spacecraft of sufficient capability to reach an extrasolar destination.

Thus there is no valid, watertight, perfect solution to the wait calculation, which means that the issue of when to leave is in the nature of a formally undecidable proposition.

In their comprehensive study, "World Ships: Feasibility and Rationale," authors Andreas Hein and colleagues conclude that "due to the large amount of resources a world ship would require, its development is likely to start after the year 2300, assuming current rates of economic growth. It is likely that at that point, alternative modes of crewed interstellar travel are already available, which might render world ships obsolete."

"Obsolete." Which is perhaps as good an epitaph as any for the troubled and problematic concept of a world ship.

7 HAIL MARY PROPULSION SYSTEMS, INC.

In American football, a Hail Mary pass is a very long forward pass typically made in last-ditch, desperation circumstances, often with mere seconds left to play, and as such has an exceptionally small chance of success, meaning ending in a touchdown. The term invokes the Hail Mary prayer in hopes of divine intervention. The completion of a Hail Mary pass is not impossible, just highly unlikely. In 2021 the sports network ESPN estimated that not even 10 percent of Hail Mary passes are successful, and reported that between 2009 and 2020 there were 193 such attempts while only 16 produced touchdowns.

And so it goes similarly for many propulsion systems designed for interstellar spacecraft: they, too, are last-ditch, desperation schemes with very small chances of actually working as promised. The decidedly iffy status of some of the propulsion concepts so far discussed – the Alcubierre Drive and Sonny White's warp drive – have led some star travel proponents to conceive of other exotic, "alternative," or overly imaginative propulsion methodologies: flying through wormholes, for example, or crackpot faster-than-light schemes such as tachyon drives. But those concepts are so far-out and unlikely as to be well beyond even Hail Mary desperation status. There are some further theoretically possible systems, however, that just might work. The least implausible of them all is the controlled nuclear fusion drive. It was this type of engine that would supposedly propel the otherwise unworkable Bussard Interstellar Ramjet as well as the second stage of the Project Daedalus starship. In its favor is the fact that nuclear fusion is the single Hail Mary propulsion technology that is currently under active development.

The standard quip pertaining to controlled nuclear fusion is that it's always been just 20 years away – and always will be. Insofar as the jest pertains to an operational nuclear fusion reactor that produces actual, usable energy, that remark still holds true: there are as yet no such reactors. But why there aren't requires some explanation.

Controlled nuclear fusion has been viewed by its scientific and commercial promoters, as well as by members of the mainstream media, and some politicians, as a "perfect" technology for producing electrical energy. It is routinely touted by the press and fusion engineers as being "inexhaustible, cheap, clean, and radiation-free." For some journalists there seem to be no bounds to the extremes of hyperbole that can be unleashed concerning the prospect of nuclear fusion. A 2023 *Wired* magazine piece about the technology ran under the headline: "It's Time to Fall in Love With Nuclear Fusion – Again," plus the subheading: "Let's indulge: Once fusion arrives, handmade suns could wipe out all human problems in a go." Really?

Well, not quite. As we have seen, the first example of nuclear fusion was the detonation of the "Ivy Mike" thermonuclear bomb that wiped out the Pacific island Elugelab on November 1, 1952. That was an uncontrolled, sudden, and explosive nuclear fusion event, and it required the detonation of a conventional fission bomb inside the device to ignite the reaction.

In nature, nuclear fusion takes place at the core of the Sun and other stars, where extremely high temperatures and pressures force hydrogen nuclei to fuse into helium atoms, releasing intense heat and light in the process. In the laboratory, scientists have tried to replicate these reactions, using hydrogen nuclei, or a deuterium and tritium combination, or a mixture of other light elements. Getting the reaction going requires the attainment of very high temperatures, in the millions of degrees Celsius, to create the hot plasma in which thermonuclear fusion can occur. And this in turn requires the input of enormous amounts of energy.

But the whole point of controlled nuclear fusion research is to reach the point where we extract more energy from the fusion reaction than was used to create the reaction itself. So far, scientists have not come close to making this happen, and there is no guarantee that they ever will.

Researchers have been trying to create controlled nuclear fusion reactions for more than 60 years. The first machine to achieve controlled thermonuclear fusion was a device using deuterium at the Los Alamos National Laboratory, in 1958. The machine, called Scylla I, reached a temperature of about 15 million degrees Celsius, and the input of a brief but intense pulse of electrical energy produced a release of neutrons, protons, and tritons that the scientists interpreted as evidence that nuclear fusion had occurred. But the hot plasma lasted for only a few millionths of a second, and there was no evidence that the reaction reached anywhere near fusion energy "breakeven," where the fusion energy produced during a pulse is equal to the energy applied from external sources to heat the plasma and get the fusion reaction going. Much less did it produce "ignition," in which the net energy output is greater than the external energy input.

Later, the United States and other countries, including the United Kingdom, France, the Soviet Union, and Korea, tested a variety of devices in an attempt to produce nuclear fusion reactions, but none of them was able to achieve a net energy gain. One of the problems in controlled nuclear fusion experiments is that the plasma gets so hot that no physical container can confine the fusion plasma: any physical vessel would be vaporized by the heat of the reaction. Instead, two different types of approaches have been developed for confining the plasma. One is magnetic confinement fusion (MCF), where the hot fusion products are confined by strong magnetic fields. The other is so-called inertial confinement fusion (ICF), where laser or particle beams are used to compress and heat a tiny capsule of fusion fuel to generate a micro-explosion of microsecond duration. Both methods have in common the fact that they consume tremendous amounts of energy.

In 2022, after 64 years of controlled fusion research, scientists using the inertial confinement method finally produced a fusion reaction in which they claimed there was a net energy gain. The event occurred at the National Ignition Facility at Lawrence Livermore National Laboratory, in Livermore, California, on December 5, 2022.

# # #

The National Ignition Facility (NIF) is the largest and most powerful inertial confinement facility ever built. Construction began in 1997, was completed in 2001, and the facility became operational in 2009. The project was five years behind schedule and cost about $5 billion, which was almost four times higher than had been estimated originally. Physically, the place was immense: a 10-story laser complex the length of three football fields. The basic idea governing it was to direct powerful laser beams at a small target containing a few milligrams of fusion fuel – deuterium and tritium – compress it to extreme pressures and heat it to a temperature at which nuclear fusion could take place.

At the beginning, scientists had the goal of reaching ignition by the end of 2012, but they consistently failed to reach that goal for more than a decade. In the years following their first experiments in March 2012, the NIF researchers increased the power of the laser beams, tested different target designs, and investigated the effects of larger and thicker fuel capsules made of materials such beryllium and plastic, as well as the use of magnetized targets, and made other tweaks to the system. The energy falling on the target capsule also grew, from 1.9 MJ (megajoules) in March 2012 to 2.15 MJ in May 2018. The energy yields of the fusion products also increased incrementally, but never reached actual ignition until the December 5, 2022 event.

In that experiment, the target assembly consisted of a small, gold cylindrical housing called a hohlraum (German for hollow room, or cavity), containing within it a smaller, peppercorn-size, spherical, aluminum capsule holding the fusion fuel. The hohlraum itself was about the size of a pencil eraser, and the exterior surface of the capsule inside it was coated with industrial grade diamond.

Prior to the laser pulse, the fuel capsule had been refrigerated to a temperature of 18 kelvin (–427 °F). The pulse itself was a product of 192 laser beams aimed at the hohlraum, and it lasted for all of 20 billionths of a second. At the moment when the lasers hit the hohlraum, the light beams generated a bath of X-rays inside it that ablated, or blew away, the capsule's diamond surface layer while at the same time compressing the frozen fuel to a pressure of more than 100 billion atmospheres. Those actions raised its temperature to 100 million kelvin (180 million °F), more than six times hotter than at the core of the Sun. The reaction was essentially a sudden implosion that moved inward at the rate of more than 100 kilometers per second.

Inside that tiny, overpressurized inferno, a number of things happened instantly and simultaneously. The tritium and deuterium nuclei fused together to form helium nuclei, also known as alpha particles. The new particles had a slightly smaller mass than the combined masses of the tritium and deuterium nuclei that produced it, and the difference in mass was released as energy according to Einstein's well-known formula that equated mass and energy, $E = mc^2$. The energy was released in the form of alpha particles, high-energy neutrons, and electromagnetic radiation.

In the December 5, 2022 experiment, for the first time ever, more energy was released in the reaction than had been put in to create it: 2.05 MJ of energy was directed to the target, which produced an output of 3.15 MJ of nuclear fusion energy. That energy increase more than surpassed the threshold level for ignition.

The results were announced by NIF officials at a news conference at the Department of Energy (DOE) headquarters in Washington, DC the following week. On December 13, DOE Secretary Jennifer Granholm said: "It's the first time it has ever been done in a laboratory, anywhere in the world. Simply put, this is one of the most impressive scientific feats of the 21st century."

But as true as that was, when *Science* magazine reported the results online that same day, in a story titled, "With Historic Explosion, a Fusion Breakthrough," the piece also ran with a curious sub-headline: "National Ignition Facility achieves net energy 'gain' with laser-powered approach." Why the quotation marks around the word "gain"? Hadn't the gain actually been achieved?

Yes and no. A paragraph later in the story, written by science correspondent Daniel Clery, explained, "If gain meant producing more output energy than input electricity, ... NIF fell far short. Its lasers are inefficient, requiring hundreds of megajoules of electricity to produce the 2 MJ of laser light and 3 MJ of fusion energy."

There was a net energy "gain" so long as the only energy flow you considered was the amount falling upon the hohlraum (2.05 MJ) and the amount released by the fusion reaction (3.15 MJ). That difference of 1.1 MJ arguably represented a net energy gain. But that's ignoring the much greater energy input that had been used to fire the lasers, which was on the order of 300 MJ, making for an overall net energy *loss* of 296.85 MJ.

The whole matter was depicted with exceptional clarity in a graphic run by the journal *Physics Today* that appeared online the same day as the *Science* piece. A block of 300 dots on the lefthand side of the image represented the 300 MJ of energy used to fire the 192 laser beams. The two little dots in the middle were the 2 MJ falling upon the hohlraum. And the three little dots on the righthand side were the 3 MJ of energy produced by the thermal fusion rection (Figure 7.1).

Matters were made even worse when you considered what it would take for the fusion reaction (which itself had lasted only a few 100 trillionths of a second) to be scaled up to the point where it could be used to generate electricity in a power station. As the *Science* piece went on to explain, "A power plant based on NIF would need to raise the repetition rate from one shot per day to about 10 per second. One million capsules a day would need to be made, filled, positioned, blasted, and cleared away – a huge engineering challenge."

NIF's ignition achievement in perspective

Energy in megajoules 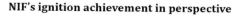 = 1

Energy required from the grid   Energy of laser fired upon hohlraum   Energy produced via fusion

Figure 7.1 NIF's controlled fusion net energy loss. Reproduced from Physics Today, with the permission of the American Institute of Physics.

In fact, the full situation was even more dire than that according to fusion energy critic Daniel Jassby, a retired research physicist who had worked for 25 years at the Princeton Plasma Physics Laboratory. In his piece "On the Laser-Fusion Milestone," published in the online journal *Inference: International Review of Science* (whose editor in chief is Nobel laureate Sheldon Glashow), Jassby concedes that the NIF results constituted a significant milestone: "For the first time, researchers have demonstrated definitively the scientific feasibility of at least one approach to controlled nuclear fusion energy." That said, the question was whether those results could form the basis for a power-producing reactor in the near future. And Jassby's answer to this was: "There is no chance whatsoever that such a power plant will appear anytime soon because a host of barely existent technologies must first be developed – or in some cases *invented* – before a fusion reactor that produces net electric power can be realized."

Jassby enumerated several reasons for that claim, noting first of all that the NIF's laser system has an electrical efficiency of about 0.5 percent and can fire full-energy pulses only once or twice a day. Second, "The tiny and intricate target assemblies comprising the hohlraum and fuel capsule cost more than US$10,000 each, require several weeks to fabricate, and hours of careful adjustments to position precisely in the target chamber." Whereas, by contrast, a successful, working implosion driver had to deliver 5 MJ of energy to the hohlraum, with an overall electrical efficiency of at least 10 percent, and be capable of firing several times per second. Most damaging of all, "A half-million identical fuel targets must be fabricated daily at a cost of no more than 20 cents each." Nobody knew how to do that.

What all of this meant was that while the NIF milestone was a proof-of-concept that controlled nuclear fusion can be achieved, nevertheless, by itself that breakthrough did not bring us anywhere near the goal of routine generation of electrical power. Nor did it meaningfully advance the prospect of a fusion-powered interstellar spacecraft.

More than seven months later, on July 30, 2023, the NIF repeated the experiment, this time attaining an output of 3.88 MJ, which was 20 percent higher than that reached previously. The *Physics Today* report on the shot said: "The fusion yields of both ignition-achieving experiments were puny compared with the 300 MJ of electricity needed to power the NIF laser." So far, this Hail Mary pass has failed.

# # #

Still, the inertial confinement method used by the NIF is only one way by which controlled nuclear fusion could be achieved. Another method was by the use of magnetic confinement, and a large group of international researchers was working on a megaproject in Europe to realize that possibility. The project is called ITER, an acronym for *International Thermonuclear Experimental Reactor*, which is located near the town of Saint Paul-lez-Durance in southern France. (ITER also means "The Way" in Latin.)

If the National Ignition Facility is immense, ITER is gigantic, stupendous, overwhelming. ITER's fusion reactor is called a tokamak (which itself comes from the Russian acronym for "toroidal chamber with magnetic coils"). The ITER tokamak has been called "the most complex machine ever built," although it is still very far from completion. According to the project's website (iter.org), the tokamak reactor will contain "One million components, ten million parts." Which is also 10 million opportunities for things to go wrong.

The theory was that the device would work by confining the fusion reaction within a magnetic field inside a doughnut-shaped vacuum chamber. Inside it, under great extremes of heat and pressure, the gaseous deuterium–tritium fuel becomes a plasma, an environment in which the D–T atoms fuse together to form a new element that was not there before, helium, and also releases energy in the form of intense heat, gamma rays, X-rays, and a flux of neutrons propelled away in all directions. The reaction is like a cloud of atom-size H-bombs exploding continuously. The hot plasma essentially replicates the conditions at the core of stars, and for this reason a 2014 *New Yorker* piece about ITER was entitled "A Star in a Bottle," a magnetic bottle.

When complete, ITER's main building housing the tokamak will stand a hundred feet tall and the tokamak itself will weigh 23,000 tons – more than twice the weight of the Eiffel Tower. The central building will be surrounded by some 40 other buildings, and collectively they will contain some 6,000 miles of electrical wiring. (Imagine all this inside a spacecraft!)

ITER is a collaborative project involving 35 member nations. "Taken together," its website reports, "the ITER Members represent three continents, over 40 languages, half the world's population and 85 percent of global domestic product." As such, it is the largest scientific research collaboration in history.

Which is perhaps one of the reasons why it has been perpetually behind schedule and over budget. When the project was first conceived, in 1993, scientists thought that the machine would be complete by 2010. As it turned out, land clearing began only in 2007, foundations for the tokamak were not laid until the years 2010 through 2014, and machine assembly did not start until 2020. And, once put together, some components were found to be defective. On the ITER website, a November 2022 "Update" reported that:

defects have been identified in two key tokamak components – the thermal shields and the vacuum vessel sectors. These defects affect the vacuum vessel sector module already installed in the pit, which has been removed for repair, as well as other delivered components. ... A proposal for an updated Project Baseline (schedule and cost) will be made to the ITER Council after the completion of a comprehensive assessment.

Nobody seems to know when, if ever, the tokamak will be up and running. And for all of its size, complexity, and cost, ITER's goal is categorically *not* to produce electrical energy. As its website admits quite openly, "ITER will not convert the heating power it produces as electricity, but [rather] it will prepare the way for the machines that can." That is, ITER is an *experiment* whose sole purpose is to show that large-scale controlled nuclear fusion is possible.

Nevertheless, a *second* proof of concept (after the NIF success) would not bring us to the advent of "inexhaustible, cheap, and clean" nuclear fusion power plants, nor to fusion-powered interstellar spacecraft. Even so fervent an interstellar travel advocate as Paul Gilster, of the website *Centauri Dreams*, did not think that fusion-powered rockets were just a step away.

"Naturally, I celebrate the NIF's accomplishment of producing more energy than was initially put into the fusion experiment," Gilster told website Space.com in December 2022. "Where we go as this evolves, and this seems to be several decades away, is toward actual fusion power plants here on Earth. But as to space exploration, we then have to consider how to reduce working fusion [reactors] into something that can fit the size and weight constraints of a spacecraft. This work is heartening, then, but it should not diminish our research into alternatives."

Let us, then, consider some alternatives.

# # #

Across the short history of interstellar propulsion theorizing, no more outlandish a technology has ever been proposed than that of antimatter propulsion. It is outlandish in the literal sense that most of the known universe is thought to be made of ordinary, normal, and conventional matter. Certainly, everything around us is made of such matter. Antimatter, to the extent that it exists at all, is thought to exist only rarely in nature: in cosmic ray collisions, for example, and in some types of radioactive decays. And even when and where a particle of antimatter does exist, it usually lasts for only a very short time, because when it comes into contact with a particle of ordinary matter, both particles are annihilated in a quick burst of energy. To emphasize how complete their mutual destruction is, it is sometimes said that the combined particles liberate "200 percent" of their mass as energy, each of them giving up 100 percent of their mass in that way.

The only place where antimatter exists permanently, or nearly so, is in the realm of science fiction. And so, in the TV show *Star Trek*, the starship *Enterprise* is of course powered by antimatter, although we are never told how the stuff is produced, stored, or directed for use as thrust.

It is at least slightly anomalous that something as rare, evanescent, and fleeting as antimatter should be proposed as a means of propelling massive interstellar vehicles through light years of space. Nevertheless, serious scientists have suggested using antimatter propulsion for interstellar travel. As we have seen, Gerard K. O'Neill proposed the use of antimatter for trips to the stars, although he conceded that such a propulsion system was "well beyond the limits of present-day technology." And in his 1988 book *Future Magic*, physicist Robert L. Forward claimed that "a starship using antimatter could travel to the nearest stars in a human lifetime," but the book gave no hint about how to produce it, confine it, or direct it. And in 1999, the American Institute of Aeronautics and Astronautics published a paper, "Antimatter Production for Near-Term Propulsion Applications," in which the authors, from the NASA Marshall Spaceflight Center and Pennsylvania State University, offered no less than six different propulsion concepts that were based on the use of antimatter as a constituent part of an overall system.

The particles of which antimatter is composed are the same particles that make up ordinary matter – protons, neutrons, and electrons – but with all of their physical properties reversed: charge (if any), spin, quantum number, and others (such as baryon number). The antiparticle of an electron, a positron, has the same mass as an electron, but carries a positive charge. The antiparticle of a proton is an antiproton but with a negative instead of a positive charge. Neutrons have no charge, nor does an antineutron, but all of its other properties have a magnitude equal to that of a neutron but of opposite sign. Antimatter is sometimes called "mirror matter" to capture the notion of reversed attributes.

The concept of antimatter is very much a child of the twentieth century. Its possible existence was first proposed by the British physicist Paul Dirac in papers published in *Proceedings of the Royal Society* between 1928 and 1931. In them, Dirac noticed that certain equations that combined quantum theory with special relativity to describe the behavior of an electron moving at relativistic speeds allowed for two solutions: one for an electron as we normally think of it, with a negative charge, and another for a positively charged particle of the same mass. Dirac interpreted this result as indicating that there must be two different particles: a normally charged electron and an antielectron, or "positron," with a positive charge. As he described it later, "the positron is just a mirror-image of the electron, having exactly the same mass and opposite charge."

But as it stood, that was just a theory. The particle in question, the mirror-image electron, was discovered soon enough, in 1932, by physicist Carl Anderson, working at Caltech. At the time, Anderson was studying the behavior of cosmic ray particles in a cloud chamber. Particles of different types, he knew, left characteristic tracks as they passed through the interior of the instrument. In a series of photographs of such tracks he came across some that looked exactly like electron tracks, except that their degree of curvature was opposite that of the curvature normally left by an electron.

At first he thought that the deflected particle might be a proton, but ruled out that possibility on the ground that a proton is much heavier than an electron, and would have made a different track from the one made by the new particle. Anderson ultimately decided that the particle he observed was an antielectron, or positive electron. He published his conclusions in "The Positive Electron," published in *Physical Review* of March 1933.

Both Dirac's prediction of positrons and Anderson's actual discovery of them were important advances in physics, and both scientists were awarded a Nobel Prize for their respective achievements: Dirac's in 1933, and Anderson's in 1936.

Still, that left the antiproton and the antineutron waiting in the wings. The antiproton was found in 1955 by Owen Chamberlain and Emilio Segré at the University of California's Bevatron, a particle accelerator at Berkeley. The new particle was identical in every way to the proton, except that its electric charge was negative instead of positive. For their discovery of the antiproton, the two scientists jointly received the 1959 Nobel Prize in physics.

The antineutron, finally, was discovered just a year later, in 1956, at the same facility, by a team led by Bruce Cork. And with that, the detection of the principal components of antimatter was complete. In 1972, the physicist Werner Heisenberg summed up the importance of these events by stating: "I think that the discovery of antimatter was perhaps the biggest jump of all the big jumps in physics in our century." And why not? It was, after all, a wholly new and unexpected form of matter.

But how could anyone hope to make an interstellar propulsion system out of particles that appeared only fleetingly in cloud chambers and in the inner recesses of immense and heavy particle accelerators?

# # #

The answer was, the stuff had to be produced artificially, in gigantic and expensive machines that essentially forced antiparticles into existence as if against their will. And indeed, when such particles were in fact produced, they existed only in minute amounts and disappeared from view forever in far less than a second. Plainly, inventing a working antimatter drive powerful enough to propel an interstellar ark was going to be a difficult undertaking.

The bulk of the antimatter that has ever been produced artificially has been made at CERN, the European Organization for Nuclear Research, near Geneva, Switzerland. "Bulk" is not exactly the right word, however, because antimatter is hardly produced in bulk quantities. In fact, the tiny quantities of it that have been produced at such great cost in terms of machinery and energy have made antimatter the most expensive material ever to exist.

At CERN, antimatter has been made in the Large Hadron Collider (LHC), which at 27 kilometers (17 miles) in length, took a decade to build at a cost of about $4.75 billion. In the LHC, protons collide with nuclei inside a metal cylinder target. Approximately four proton–antiproton pairs are produced for every four million collisions. Left to themselves, the antiprotons would be annihilated by contact with the inner walls of the collider, and so to avoid that outcome the antiprotons are separated using magnetic fields and are shunted to the facility's Antiproton Decelerator, where they are slowed from 90 percent to 10 percent of the speed of light. Finally, the antiprotons are confined inside a device called a Penning trap, in which strong magnetic fields hold the particles in an ultrahigh vacuum. In 1978, CERN produced several hundred antiprotons and kept them circulating for a period of about 85 hours. According to a CERN press release, "Antimatter, in the form of antiprotons, has been stored for the first time in history."

But antiprotons are only subatomic particles – *parts* of atoms. Producing antimatter in the form of whole atoms is a more difficult proposition, and storing them for a period of time is even more challenging. In 1995, CERN reported the production of 11 antihydrogen atoms. In "The Story of Antimatter" CERN's website described the process:

A team led by Walter Oelert created atoms of antihydrogen for the first time at CERN's Low Energy Antiproton Ring (LEAR) facility. Nine of these atoms were produced in collisions between antiprotons and xenon atoms over a period of 3 weeks. Each one remained in existence for about 40 billionths of a second, travelled at nearly the speed of light over a path of 10 metres and then annihilated with ordinary matter. The annihilation produced the signal that showed that the anti-atoms had been created. This was the first time that antimatter particles had been brought together to make complete atoms.

It was not until 2010 that a team of CERN scientists not only produced a countable number of antiatoms but also brought them to a complete halt in a magnetic trap – albeit not for very long. As reported by Eugene Reich in *Nature*, a CERN group calling itself the ALPHA collaboration (for Antihydrogen Laser Physics Apparatus), "has managed, 38 times, to confine single antihydrogen atoms in a magnetic trap for more than 170 milliseconds."

By any standard, 170 milliseconds was not a long time. Nevertheless, "we're ecstatic," Jeffrey Hangst, the ALPHA spokesman, said of that result. "This is five years of hard work." And to trap those 38 atoms, one by one, the scientists had to run the experiment 335 times. "This was ten thousand times more difficult" than creating untrapped antihydrogen atoms, Hangst said.

The very next year, in 2011, the same ALPHA group at CERN managed to confine antihydrogen atoms for 1,000 seconds, or a little more than 16 minutes. At about this time CERN released a statement putting these results into perspective:

Even if CERN used its accelerators only for making antimatter, it could produce no more than about 1 billionth of a gram per year. To make 1 g of antimatter ... would therefore take about a billion years. The total amount of antimatter produced in CERN's history is less than 10 nanograms – containing only enough energy to power a 60 W light bulb for 4 hours.

Later researchers at CERN, Fermilab, and at the Lawrence Livermore Laboratory have reported the creation of billions of antiparticles, and the maintenance of progressively longer storage times. However, even these quantities amount to merely nanograms of antimatter, and were produced solely for research purposes.

# # #

There are two substantial problems with using antimatter as the power source of an interstellar propulsion system. The first is that of making enough antimatter to constitute a system that could propel a massive interstellar spacecraft to the stars. Given the extremely low output rate of existing antimatter production facilities, this is the biggest obstacle to be overcome, and perhaps is even an insurmountable one for the foreseeable future. A recent (2022) technical article, "Antimatter Propulsion and Its Application for Interstellar Travel: A Review," by physicist Shrijan Sharma, says that, "The most significant hurdle faced by the idea of antimatter propulsion is the efficient production of antimatter."

Worse, Sharma finds no realistic and cost-effective method of overcoming that hurdle. "Extensive research would be required in the field of production and storage, and new technologies would need to be invented. A lot of research would also be required in designing an efficient rocket that would be able to function on antimatter propulsion and utilize the products of annihilation successfully." In the end, the author concluded that "antimatter propulsion is not a viable option for interstellar travel in the next few decades."

The second problem is that since antimatter destroys conventional matter, it is impossible to store it in normal physical containers; and for the same reason it would be impossible to direct antimatter thrust by means of conventional, physical exhaust nozzles such as are used in chemical rocketry.

"The problem with working with antimatter is that one has to keep it from touching matter," says James Annis, a physicist at Fermilab's Center for Particle Astrophysics. "And that is surprisingly hard to do." Immense and powerful magnetic fields are required to accomplish that goal, and the more antimatter there is in storage, the stronger the magnetic field needs to be. The magnetic field must also never fail: *any* interruption in the supply of electricity, any power outage or little blip, no matter how momentary, would cause an explosion rivaling that of a nuclear bomb. Veritably, the antimatter starship is the *Hindenburg* of interstellar spacecraft.

It is incongruous, given the seriousness of these problems, that one of the first antimatter rocket concepts had been proposed early on, when the only type of antimatter then known was the positron. This was Eugen Sänger's so-called "Photon Rocket," which he described in 1953. Positron and electron pairs would be metered out of a storage container, after which they would mutually annihilate, creating a shower of gamma rays. The problem was that the gamma rays would scatter in random directions unless there was some way of channeling them into a directed exhaust stream. Sänger was not successful in coming up with a scheme that would work, and so he ended up having effectively designed a gamma ray *bomb* rather than an antimatter rocket. Gamma rays (which are photons) are highly energetic particles, they penetrate all known materials, and would be lethal to a starship crew.

Taken together, the formidable problems with antimatter production and storage explain why, although there are lots of imaginary antimatter starship designs on paper, and in the pages of *JBIS*, in YouTube videos, in science fiction stories, TV shows, artworks, and movies, no government agency or private aerospace company is currently building an actual, real-world antimatter rocket. There are plans, ideas, designs, but no hands-on, tangible rockets in the works.

There is, however, a possible exception in the form of a company called Hbar Technologies, LLC. The company was created in 2002, in Chicago, by Gerald Jackson, a former Fermilab particle accelerator physicist, and partner Steven Howe, formerly a physicist at the Los Alamos National Laboratory and principal investigator on a short-lived NASA study, "Antimatter Driven Sail for Deep Space Missions" (2004–2005). The first word of the company's name, "Hbar," is a spelling-out of the physicist's symbol for an antiparticle, which was a letter, such as h for hydrogen, with a bar over the top of it, $\bar{h}$, and pronounced "H-bar." The company's motto was: "Making Antimatter Matter."

The idea of Jackson and Howe was to do antimatter research on their own, privately. They had a plan by which they needed a small amount of antihydrogen to use in order to propel an unmanned probe to Alpha Centauri at 10 percent the speed of light. Supposedly, this would get the craft there in only 40 years.

Their proposed spacecraft consisted of a 5-meter diameter carbon-fiber sail coated with a layer of uranium. The craft's fuel would be antihydrogen frozen into pellets. The plan was to stream the antihydrogen pellets toward the sail, whereupon they would induce a fission reaction in the uranium, propelling the spacecraft forward.

With NASA funding having dried up, Jackson and Howe decided to raise money via a crowdfunded Kickstarter campaign. This, they hoped, would bring forth a groundswell of support that would allow them to optimize their design and see it through an initial round of testing. They were looking for only $200,000 – enough to get started. "We will then need funding on the order of $100 million to actually build small prototype propulsion and power systems," Gerald Jackson explained. "But you're talking billions to send spacecraft out to real destinations."

Hbar launched its Kickstarter effort in mid-2016 and hoped for the best.

Money did come in, albeit slowly. A year later, they had accumulated a total of 62 backers who collectively pledged $2,280 (for an average contribution of $37) to help bring the project to life. This was somewhat less than planned.

Jackson chose to reward the top 20 contributors with a commemorative token of their participation. This turned out to be a 4-inch-diameter, iron-on shoulder patch. The patch, in full color, showed a rendering of the proposed spacecraft, with depictions of various antiatoms around the border, and with the legend: "Antimatter Fuel Production," and "2016–2017 Kickstarter."

It is easy to imagine one or more of the lucky recipients saying to themselves: "I hoped to get us launched toward the stars, and all I got was this lousy shoulder patch." That, anyway, was the end of the project.

# # #

And then, finally, there is the "rocketless rocket," the ultimate Hail Mary machine. This is a spacecraft that carries no propellant, lacks an engine for propulsion, and can operate continuously without refueling. Instead, the vehicle flies through space by being pushed from afar, by a high-energy, powerful laser beam sent from within the solar system. In the form in which it is most commonly discussed today, this concept was conceived by the American physicist Robert L. "Bob" Forward.

Forward worked at the Hughes Research Laboratories (now HRL Laboratories), which is a sprawling research complex located high on a hill overlooking Malibu, California, and which commands a lordly view of the Pacific in the distance. The entire setting was highly conducive to the birth and development of great ideas and new technologies. One of them was the laser. For it was here at Hughes that in 1960 the engineer and physicist Theodor H. Maiman invented the first operational laser device. The laser, which was an acronym for "light amplification by stimulated emission of radiation," amplified and intensified a beam of monochromatic light whose rays were coherent, meaning that they were in phase with each other and so radiated out in the form of almost perfectly parallel waves. Ordinary white light, by contrast, is a combination of all the colors of the visible spectrum, and because its rays are out of phase with each other they tend to disperse and spread out laterally in all directions.

Two years later, in 1962, and also at Hughes, Bob Forward had the idea of using a laser beam to propel a lightsail across interstellar distances. In retrospect, this was an "obvious" use of the new laser technology. The idea of lightsails propelled by sunlight had been proposed as far back as 1924 by the Russian space theorist Friedrich Tsander, and perhaps even earlier by Tsiolkovsky. Light has momentum, and enough sunlight falling on a large enough sail could drive it from Earth orbit to essentially any other point within the solar system.

The advantages of using laser beams to do the same thing were twofold. One, since laser light was coherent and did not spread out, a beam could traverse thousands of miles while remaining essentially as concentrated as it had been at the beginning. Two, laser light was vastly more powerful than ordinary light; indeed, a laser beam is at least 10,000 times more powerful than sunlight. These advantages meant that it might be possible to use a laser to propel a spacecraft across even interstellar distances.

When Bob Forward first conceived of the concept, he thought that such a use of lasers would be confined to one-way, flyby missions since the only thing a laser beam could do was to keep pushing the lightsail away from the laser transmitter within the solar system. There was no way he could think of by which a pusher beam could stop the lightsail, or even slow it down near the target star. Even if the laser source stopped emitting light entirely, the lightsail would keep flying off into the great beyond by dint of its own momentum.

Taking the concept further, Forward saw that the sail for even an unmanned, one-way, flyby mission would have to be very large. Assuming a total, combined mass of both sail and an associated probe module to be 1,000 kg (2,204 lb), Forward calculated that the sail would have to be 3.6 km (2.25 miles) in diameter. And the power needed to push the probe to the nearest star would be on the order of 65 gigawatts.

"While this is a great deal of power," he later wrote, "it is well within future capabilities. Higher power levels than this are generated by the Space Shuttle rockets at liftoff."

Still, Forward was irked by the fact that there was no way of using the laser beam to stop the ship or slow it down. This really bothered him, and he thought about the problem for a solid 20 years, circling back to it again and again in his mind. And then, finally, he hit upon a solution.

"While trying to find a new way of travelling to the stars for a novel I was writing, I realized that if the lightsail were separated into two parts, then one part might be used as a mirror to reflect the laser light back toward the solar system," he explained later. "The retrodirected light might then be used to decelerate the other portion of the lightsail. When I worked out the equations and put numbers into it, I found that not only was it a good science fiction idea, but it would really work."

Forward went through another set of calculations for an interstellar mission in which a larger, unmanned spacecraft would rendezvous at the target star and explore it by robots in some detail. The sail for such a mission would have a diameter of 100 km (62 miles), have a mass of 785 metric tons, and would accelerate to a speed of 10 percent the speed of light. He also calculated how much power it would take to accomplish this feat, and found that the answer was 7.2 terawatts.

"It should be noted at this point that the total power output of the entire world is about 1 TW," he wrote. "This amount of laser power is not trivial and will require a significant commitment to build a large array of solar-powered lasers in space."

When he was done with fully working out all of these systems, specifications, and schemes, Forward penned his magnum opus, "Roundtrip Interstellar Travel Using Laser-Pushed Lightsails." It was published in the *Journal of Spacecraft and Rockets* in 1984.

Forward also came up with a way of making a roundtrip interstellar voyage by separating the lightsail into *three* parts, and using energy reflected from the third part to push the spacecraft back to where it had come from. This would be a crewed mission of truly astounding size, with a lightsail at launch of 1000 km (621 miles) in diameter, and a total, combined mass of 78,500 metric tons. In the published paper mentioned above, Forward says that, "if we want to provide a constant acceleration, the laser power would have to be increased from 43,000 TW at the start to 75,000 TW or more at the end of the acceleration phase."

We must pause here to note that if, as Forward says, the power output of the entire world is 1 TW, then the 75,000 TW as quoted just above is equal to 75,000 times the total power output of the entire world. But Forward does not say how such an absurdly high level of power is to be generated nor whether it would be possible, even in principle, to do so. We are forced to conclude, then, that as he describes it, this particular spacecraft and the power level required to propel it across interstellar distances lie more in the realm of fantasy than fact. In addition to his paper in the *Journal of Spacecraft and Rockets*, Forward also described these concepts in his 1988 book, *Future Magic: How Today's Science Fiction Will Become Tomorrow's Reality*. It may be, however, that that the laser-pushed lightsail spacecraft that he describes both in his 1984 paper and, later, in his book, really are science fiction, and not reality, whether tomorrow or ever. The scheme is devoid of plain horse sense.

# # #

Forward himself acknowledged that there were some unknowns, loose ends, and open questions regarding his own setup. "First, there are the engineering problems of scaling the present lasers to higher power levels, then making large numbers of these lasers operate as a coherent phase-locked array," he wrote in his journal article. "Second, there is the political problem of making the decision to spend technological resources for interstellar flight, rather than for some other goal."

Other theorists have seen further social difficulties. Eugene Mallove and Gregory Matloff, early critics of the idea, have said: "Beamed power for starflight carries a potential *disadvantage*: the need to maintain economic and political stability on the homefront during a possibly very long acceleration period and perhaps during deceleration as well."

Another problem with the idea of beamed propulsion stems from the fact that that it is one of the few cases in the interstellar literature in which the spacecraft is physically decoupled from the system propelling it. This raises the problem of tracking an ever-receding spacecraft and aiming the beam accurately so that it reaches the vehicle on target. It must also be recognized that the spacecraft might be deflected from its intended path by encounters with clouds of interstellar gas and dust, or pushed off course by unsuspected magnetic fields, or by other unseen factors that might draw it off its preplanned track.

Further, the problems created by the decoupling of the ship from its propulsion system become progressively greater with distance traveled, and are especially acute over distances measured in terms of light years. Corrective requests from the ship back to the operators of the laser array will take ever longer to be received and acted upon. It will also take progressively longer for any aiming correction to take effect upon the targeted spacecraft.

The farther away the spacecraft gets from the laser sending array, the more crucial, and more protracted in time the feedback loop becomes. In the voyage to Epsilon Eridani that Forward mentions, the target star is 10 light years away from Earth. If, for example, the crewed spacecraft notices an unintended deflection from its planned path through space while it is two years into its voyage, then it will take two years for an emergency message to be received at the laser array and another two years for the effect of any beam aiming correction to be received at the spacecraft. But within that four-year delay period the ship might have experienced a further deflection from its path to the degree that the requested aiming correction is never actually received by the ship. And if the ship makes an aiming request when it is seven years out, then any applied correction would not be received until 14 years had elapsed, by which time the 10-year-long trip would supposedly have ended.

There are the further problems of failures at either end of the decoupled systems: at the laser array in orbit around the Sun, or at the lightsail and crew module, the former of which is susceptible to micrometeorite damage.

Every one of these problems is quite substantial. While none of them is necessarily fatal to such a project in principle, nevertheless from a practical, commonsense point of view, any design that requires for its operation an amount of power that is 75,000 times the total power output of the whole world, as does Forward's laser-pushed lightsail, is a project that is doomed from the start.

8 THE FATE OF THE CREW

Let us optimistically assume that sooner or later a workable interstellar propulsion system will be found, and also be built and successfully tested in space. While this would be a great advance toward making interstellar travel possible, it nevertheless does not automatically follow that a voyage to the stars will in fact be attempted. There are a few other issues that must also be settled first: for example, a habitable exoplanet must be identified. It must be suitable for human colonization and ought to be a reachable distance away from Earth within a reasonable period of travel time. Second, engineers must provide a plausible space vehicle design architecture, and a spacecraft of that design must then be constructed and tested successfully. Such a craft does not yet exist, one among many reasons being that the specifications for it depend in turn upon the size and makeup of the likely boarding population. But both of those factors are still unknown. In addition, and perhaps most important of all, an unprecedented level of funding and resources must be allocated to the project. A multigenerational interstellar voyage will be an effort many orders of magnitude greater than past and current gigaprojects such as the Apollo program, ITER, the Large Hadron Collider, and the failed (for want of adequate funding) Superconducting Super Collider (or SSC).

There is also the problem of knowing where such funds will come from or indeed whether they will ever be provided at all, something that is made doubtful in light of the fact that the trip will be of relatively little benefit to the sponsors, and the further fact that there is no apparent sense of urgency about anyone's making the trip to begin with. Finally, it is a truism that nobody knows the future. In the face of all this, it might be thought that it is impossible to project what life aboard an interstellar spacecraft would be like, much less to know the probable fate of its crew.

But in fact matters are not as bad as all that, for it is possible to entertain a range of different outcomes and then to specify the likelihood of one or more of them happening within a given context. In this case the context is a set of features that are common to *any* ship architecture, of any size, headed anywhere. And what's in common to all crewed spacecraft missions, of whatever ship type, distance traveled, or time frame to completion, is that the craft will be crewed by people living in the extreme physical environment of space and also in a closed and isolated social environment. Biologically and psychologically, human beings are a well-known quantity. "Human nature" is fixed and will probably not change very much if at all in space. This means that we can know with at least some minimal degree of certainty what will be the likely impact on a group of humans embarked upon a long-duration space mission such as a voyage to the stars, and can hazard a best guess about what the ultimate fate of the crew might be.

First of all, even in space the crew members will be susceptible to all the same strains and stresses to which human beings are subject during the course of ordinary and everyday life on Earth. Added to those initial vulnerabilities will be the further physical and psychological burdens imposed upon the crew members by the rigors of living in close quarters with others aboard a spacecraft that is traveling through the interstellar void for a prolonged period of time.

By the year 2023, more than 560 individuals had been sent into in space, starting with Russian cosmonaut Yuri Gagarin and his 108-minute orbital flight on April 12, 1961, and stretching through to NASA astronaut Frank Rubio and his record-breaking 371 consecutive days in space aboard the International Space Station, which ended with his return to Earth on September 27, 2023. Collectively, this elite band of spacefarers has become one of the most intensively studied group of biological specimens of any kind, providing medical researchers with an ever-increasing body of knowledge about the effects of spaceflight on human physiology and psychology. And if there is one lesson to be learned from the cache of information derived from the experiences of these individuals, it is that human health is certainly not *improved* by going into space. Or, as Kelly and Zach Weinersmith put it in their book, *A City on Mars* (2023), a skeptical take on the prospects of Mars colonization, "The good news from space is that it doesn't kill you immediately."

The health consequences of going into space are illustrated by the findings of three important and authoritative scientific studies of the effects of spaceflight on human beings. The first of the three is the NASA Twins Study of 2015–2016. During the course of this experiment, the health status of two identical twin astronauts was monitored before, during, and after one twin, Scott Kelly, spent 340 consecutive days in space aboard the International Space Station, while the other twin, Mark Kelly, himself a retired astronaut, remained behind on Earth. (Scott Kelly was accompanied on the ISS by the Russian cosmonaut Mikhail Kornienko.)

Upon his return to Earth, which was not uneventful, Scott Kelly's health was continually and minutely assessed by a total of 10 different research teams from across the United States, plus another one in Germany. Altogether, these researchers had access to data about *everything* concerning the health of Scott Kelly, down to comprehensive molecular, genetic, microbial, gene-expression, and other influences relevant to Scott's time in space. In 2019, the journal *Science* published the findings of these teams in a 21-page paper (plus supplementary material and references), which had been cowritten by 77 authors: "The NASA Twins Study: A Multidimensional Analysis of a Year-Long Human Spaceflight." Given this abundant store of data about one individual, it may be true to say that Scott Kelly is the single most exhaustively health-monitored human specimen in all of history. Further, Kelly himself told his own story in his book, *Endurance: A Year in Space, a Lifetime of Discovery*, in 2017. Together, these two documents present a virtually complete account of what it is like to spend a year in space.

The findings of the *Science* team of researchers, plus Kelly's own first-person chronicle, represent a detailed narrative of continuous physical and psychological impairment and disruption, some of the damage being temporary, other harms being long-term. "Extensive multisystem changes occur in spaceflight," the team-written report said. "Differential gene expression analysis indicated that many immune-related pathways were significantly changed inflight across all cell types, including the adaptive immune system, innate immune response, and natural killer cell–mediated immunity."

There were bone density and muscular losses. In his book, Kelly reports how "my body has noticed that my bones are not needed in zero gravity. Not having to support our weight, we lose muscle as well. Sometimes I reflect that future generations may live their whole lives in space, and they won't need their bones at all."

The scientists found multiple disrupted gene sequences, especially in mitochondrial cells, which provide the energy needed to power a cell's biochemical reactions. In Scott's case there had been a total of 8,564 differentially expressed genes. There was DNA damage due to "ionizing radiation exposure during spaceflight, specifically galactic cosmic ray-induced cytogenetic damage," and "changes over time in the gastro-intestinal microbiota."

"I have been exposed to more than thirty times the radiation of a person on Earth, equivalent to about ten chest X-rays every day," Scott Kelly reported. "This exposure will increase my risk of fatal cancer for the rest of my life."

There were also "neuro-ocular changes," alterations in the cross-sectional area of the eyes, which became flattened, together with other odd effects. Kelly described what happened when he tried to go to sleep. "Even though my eyes are closed, cosmic [ray] flashes occasionally light up my field of vision, the result of radiation striking my retinas."

He experienced mental problems. "Most notably, cognitive speed decreased for all tests except for the DSST [digit symbol substitution task], and accuracy decreased for all domains except for spatial orientation postflight," the scientists said. "This postflight decline in cognitive performance persisted up to 6 months postflight in both speed and accuracy domains." For Scott Kelly, "If I'm trying to do something complex, I actually start to feel stupid, which is a troubling way to feel on a space station."

Further, "returning to Earth is an especially stressful event that represents one of the greatest physiological challenges of spaceflight," the scientists said. "In the immediate post-landing period, manifestations of this include exaggerated or impaired cardiovascular, musculoskeletal, and stress and inflammatory responses."

The morning after he landed back on Earth, Kelly was attempting to get out of bed, and ... "I feel like I'm fighting through quicksand. When I'm finally vertical, the pain in my legs is awful, and on top of that pain I feel something even more alarming: all the blood in my body is rushing to my legs, like the sensation of the blood rushing to your head when you do a headstand, but in reverse. I can feel the tissue in my legs swelling. ... Normally if I woke up feeling like this I would go to the emergency room."

There were many other changes to associated organs and systems, too numerous to detail in full, or even in part. Suffice it to say that the mental and physical deterioration that Scott Kelly experienced during his year in space would also be suffered by each member of the succession of generations to be born aboard a starship making an interstellar voyage across hundreds or even thousands of years. And these effects would be further worsened by a second factor: Kelly spent his entire time in space comfortably within the confines of Earth's protective magnetosphere, whereas interstellar travelers would lack that advantage.

The magnetosphere is a region of space surrounding the Earth in which charged particles are deflected by our planet's magnetic field. This magnetic field shields the Earth from a substantial amount of solar and cosmic particle radiation. It is part of what makes life on Earth possible. The magnetosphere extends from the ground up to 40,000 miles out into space. Its protective effect, however, weakens with altitude, and even on Earth, those who live at higher elevations receive higher levels of cosmic and solar radiation than those living at lower heights.

The International Space Station orbits at an altitude of 250 miles, well inside the magnetosphere, but Kelly was still affected by ionizing radiation. The void outside the solar system affords no such protection. An interstellar vehicle traveling through deep space would be completely outside the shielding force of Earth's magnetosphere. This means that the crew would be potentially exposed to many more times the level of cosmic ray damage than was sustained by astronauts, and be exposed for a much longer time period. Physical shielding can be built into the spacecraft, but very energetic cosmic ray particles traveling through the interstellar medium can interact with the shielding material and emit a shower of secondary particles than can damage cells and their genetic material. Providing adequate shielding is thus a major challenge for the designers of interstellar vehicles.

# # #

A limitation of the NASA Twins Study was that its findings pertained to only a single individual who might or might not be representative of the broader community of spacefarers. What of the other 560-plus men and women who have spent time in orbit?

The second canonical study here is a recent (2023) synoptic paper in the journal *Cell*: "Human Health during Space Travel: State-of-the-Art Review." This report, which drew upon a total of 289 references, was the joint product of 17 coauthors from the United States and one in the United Kingdom, who collectively represented a wide variety of health specialties. The litany of ills enumerated by this group both duplicated and added to those experienced by Scott Kelly, who proved in retrospect to be all too typical of the breed.

The astronauts in general suffered a wide range of health effects upon practically every organ and system of their bodies. Spaceflight altered cardiovascular physiology to the point that it led to cardiac atrophy. It increased the risk for both atrial fibrillation (irregular heartbeat) and atherosclerosis (plaque buildup in the arteries). Short-duration spaceflight caused gastrointestinal symptoms, while long-duration spaceflight posed an increased risk of radiation-induced gastrointestinal cancer.

Bacteria encountered in the space environment were more resistant to antibiotics and were more harmful than those encountered on Earth, while traditional prescription medications did not always function in space as intended. Weightlessness led to alterations in spatial orientation, shape recognition, and changed depth and distance perception. There were also changes to smell and taste sensations. And not only did microgravity produce structural changes to the eye while in space, those changes were later documented to persist for up to seven years of long-term follow-up testing.

Astronauts suffered weakened ability to concentrate, had short-term memory lapses, and had to cope with an inability to multitask. They coined the term "space fog'" to cover the generalized lack of focus, altered perception of time, and the cognitive losses that came with spaceflight. Despite their frequent use of sleeping pills, many of the subjects were often sleep-deprived because of the incessant and omnipresent noise of whirring, wheezing machinery.

The one organ system that remained relatively unaffected by long-duration spaceflight was the pulmonary system, for the functioning of normal human lungs was largely unchanged in space. The skin, by contrast, was highly susceptible to space radiation, and caused increased rates of basal cell and squamous cell carcinomas.

In addition to showing the need for radiation shielding, a second implication of both this study and the NASA Twins

Study for interstellar travel is that for the health of the crews across a prolonged time period, artificial gravity clearly must be provided. There are two ways of accomplishing this, the first of which is by rotating the spacecraft about an axis parallel to the plane of the living area, after the fashion of an O'Neill cylinder, creating an apparent gravitational force equal to 1 g. The other method is by accelerating the spacecraft continuously in a straight line, as has in fact often been suggested for interstellar voyages, and then decelerating at 1 g when approaching the target star system.

# # #

It is possible, at least in principle, albeit perhaps expensive and difficult in practice, to provide adequate shielding and artificial gravity to interstellar crew members. But the same cannot be said for protecting crews from the third category of space illnesses, those arising from prolonged enforced togetherness, isolation, and boredom. An important paper on this subject is: "The Burden of Space Exploration on the Mental Health of Astronauts: A Narrative Review," published in *Clinical Neuropsychiatry* in 2021. A group of seven authors assessed the role that living in the "extreme environment" of space played on the mental health of American astronauts and Russian cosmonauts across the full duration of the space programs of the two nations.

Among their findings was that "mood issues of the astronauts have long been reported, and they may also compromise the fulfillment of the mission task, as in the case of the abrupt abort of the Soyuz T14-Salyut 7 mission in 1985, which was suggested to have been partly caused by crews' depression."

Astronauts have displayed symptoms of reduced resilience, decreased drive and energy levels, and passivity. Negative interpersonal relations have caused anxiety, with the heterogeneity of the space crew in terms of size, ethnic background, and languages having resulted in tension and communication issues among crew members. Prolonged isolation and being forced to share a confined environment with the same people always around posed a risk to the mental wellbeing of those involved.

The synthetic environment of a spacecraft, whether it's orbiting the Earth or bound for the stars, provides a constant, inescapable, and unpleasant mix of excessive light, noise, vibration, bad smells, and sudden variations in ambient temperature, spacecraft internal pressurization, and air content. When the CO_2 air scrubber canisters failed to keep up with carbon dioxide buildup in the space station, Scott Kelly knew it immediately, he didn't have to consult his laptop readout. "I don't need to," he said,

> I can feel it. I can sense the levels with a high degree of accuracy based only on the symptoms I've come to know so well: headaches, congestion, burning eyes, irritability. Perhaps the most dangerous symptom is impairment of cognitive function – we have to be able to perform tasks that require a high degree of concentration and attention to detail at a moment's notice, and in an emergency, which can happen anytime, we need to be able to do those tasks right the first time. Losing a fraction of our ability to focus, make calculations, or solve problems could cost us our lives.

To relieve the stresses, strains, and rigors of the synthetic environment surrounding them, astronauts and cosmonauts hoped at least to get a good night's sleep. But this proved surprisingly difficult. One study (Arone et al., 2021) reported:

> about 75% of the crew members had taken a hypno-inducer, with drug take being reported on 52% of nights and the use of two drugs in around 17% of cases. The most frequently used medications were zolpidem and controlled-release zolpidem; others were temazepam, eszopiclone, melatonin. Another study looked at medication used during flight and found that 94% of astronauts were taking some medication, 45% for sleep disorders.

This combination of carbon dioxide intoxication, unwanted noise, the vibration of buzzing and hissing machinery, plus half-asleep, drug-addled crew members does not make for a reassuring picture of life aboard an interstellar spacecraft. "Illusions and hallucinations also frequently appeared in some reports of space missions, possibly as a result of isolation or sensory deprivation," the scientists said.

And all of these mental and physical ill effects were for relatively short-duration space missions lasting up to six months. The situation can only degrade further, and progressively, on a multigenerational interstellar journey as the years pile up, decade following decade. Added to these physiological impairments and losses will be the psychological and emotional effects stemming from a large human population being enclosed in an artificial environment traveling through deep space for hundreds of years. This means crowding, lack of privacy, protracted isolation, factionalism, trance states and depression, unforeseen emergencies, and every other source of misery and conflict that is found on the home planet, not excluding suicides and crimes of violence, including murder.

Additionally, the role of religion cannot be dismissed: after all, where humans go, so too do their faiths. During his short stay on the Moon after the landing of the Apollo 11 lunar excursion module, astronaut Buzz Aldrin administered the Eucharist to himself. As he later told a Christian periodical: "It was interesting for me to think: the very first liquid poured on the moon, and the very first food eaten there, were the communion elements."

At the 100 Year Starship public symposium in Houston in 2012, the Rev. Dr. Alvin L. Carpenter, Pastor of the First Southern Baptist Church in West Sacramento, California, gave a talk entitled, "The Non-Promise of [Bringing] Earthbound Religions into Space." In it, he said that religious beliefs aboard a starship would be a poisonous influence. Religious wars were a fact of history, and "we don't want to export this to the stars." The appearance of a charismatic leader who would sow dissent among the crew would be disastrous inside a starship. "If there's any way you want to make this trip fail, bring earthbound religions."

The problem is, there is no way to prevent it.

Finally, the very idea of a multigenerational voyage of course presupposes that human reproduction and successful procreation in space are possible, which has not been even remotely demonstrated scientifically. First, so far as is known sex among humans has never occurred in space, much less has there been a pregnancy resulting in a live birth in orbit of a healthy space baby. In fact, the amount of data that we have pertaining to the effects of microgravity and/or ionizing radiation is remarkably slim, and is derived largely from experiments on mice, rats, and fruit flies. A recent relevant study by an international group of authors is: "Effects of Space Flight on Sperm Function and Integrity," published in *Frontiers of Physiology* in 2022.

Mice have been flown into space and have been kept aboard various spacecraft, including the ISS, for periods ranging from three to seven weeks. In addition, in 2009 a group of 10 mice spent 91 days on the ISS, but only three returned to Earth alive. Their testes were examined by light microscope and showed considerable damage. "The histology showed degenerative changes," the report said. "This included disorganization and a slight reduction in the thickness of the spermatogenic cells," along with other kinds of cellular damage.

On the basis of this and other such experiments on a range of animals and time frames, the general conclusion of the authors was clear. "Exposure to microgravity and ionizing radiation can adversely affect spermatogenesis and as well alter sperm DNA/chromatin integrity. Since the mature human spermatozoa do not have the capability to repair their DNA, sperm DNA damage depending on the radiation dose may lead to permanent infertility, and/or increase the risk of congenital anomalies occurrence in the offspring."

None of this was good news for the future prospects of a multigenerational interstellar space mission.

# # #

As it happens, none of these studies of long-duration stays in space aboard the International Space Station or elsewhere constitutes a suitable analogue of what a multigenerational interstellar mission would be like if we assume that those aboard the starship would be provided with 1 g artificial gravity and would be largely shielded from cosmic radiation. And in fact another, even longer-duration project that was conducted not in space but on Earth represents a better equivalent of a crewed interstellar mission. The project, which ran during the 1990s, was very big news and a very big deal at the time, but is mostly forgotten today. It was known as Biosphere 2.

Biosphere 2 was a 3.15-acre closed ecological complex built between 1987 and 1991, and located at Oracle, Arizona, about 30 miles north of Tucson. It consisted of a set of structures designed to miniaturize and encapsulate Earth's environment in order to determine what would be the minimum total ecosystem in which a small group of eight people could live for two full years, completely on their own, without any material assistance or inputs, except for energy, from the outside world. Since the facility would be self-sufficient and hermetically sealed, it would approximate living conditions aboard an interstellar ark.

This passably far-out idea emerged from a group of visionaries living in a forward-looking and innovative compound near Santa Fe, New Mexico starting in the 1970s. It was called Synergia Ranch to reflect the theory that living systems, including the whole of Earth itself, were self-sustaining, synergistic, homeostatic environments that resembled individual organisms.

Visionary in chief at Synergia was one John P. Allen, a systems ecologist who dreamed of building the prototype version of what could eventually be a self-supporting space colony on the Moon, Mars, or a starship. Allen held a BS from the Colorado School of Mines and an MBA from the Harvard Business School. During a series of meetings at Synergia, Allen managed to gain financial backing from Ed Bass, a wealthy Texan oil heir who styled himself an "ecopreneur," a businessman who could make money from various environmentally friendly biological enterprises that could also turn a profit.

Over the course of several years, Bass provided the funding for the planning, construction, and operation of what would become Biosphere 2 (the Earth itself, of course, being Biosphere 1). It would be built by a company called Space Biosphere Ventures (SBV), of which Bass would be chairman. This would be a wholly private undertaking, with no government backing, and scant help from NASA, which was skeptical of the project.

The architect of Biosphere 2 was Phil Hawes, a Frank Lloyd Wright protégé who in the end produced plans for an appropriately futuristic, closed, space greenhouse that indeed would not look out of place on the Moon. What finally arose on the Arizona desert floor was a multifarious and convoluted set of interconnected structures that would ultimately house 3,800 species of plants, animals, birds, bees and other insects, worms, bacteria, fungi, and, not least, people (Figure 8.1). They would all live and thrive in peace and relative harmony, they hoped, demonstrating that a closed and self-sufficient mini-Earth was a realistic possibility over a substantial period of time.

Even thinking about how a workable, hermetically sealed, self-sustaining ecosystem could be designed and then realized in practice required a major period of forethought, research, and advance planning, but the Synergia group included a number of experts in systems engineering and related technologies. Plus, they would be aided in their planning by a roster of eminent

Figure 8.1 Biosphere 2. Credit: Arizona Board of Regents/The University of Arizona.

outside specialists from a number of organizations including the Smithsonian Institution, the New York Botanical Gardens, and the US Geological Survey, among others. And at length, after a series of "biome design meetings," the group came up with an overall ecological scheme that seemed plausible. Biological recycling of human and farm animal wastes, plus effluents from workshops and analytic and medical laboratories, would be provided by a two-stage system consisting of primary treatment in anaerobic holding tanks followed by circulation in specially constructed wetland areas where microbes would remove contaminants. Plant foods would be grown in an agricultural area, and animals such as pigs and their piglet litters would provide fresh meat. Goats would produce fresh milk, and fowl, fresh eggs. The general concept was so bold as to be breathtaking, and perhaps was even a little naïve. But over time, all of the miscellaneous elements would be obtained in sufficient quantities, and then deftly arranged and put together in a manner that made biological sense.

The main structure was an eight-story-high, vaulted-roof pyramidal greenhouse made up of 77,000 metal struts that held a total of 6,600 panes of glass. Below this glass covering stood the canopy of a tropical rainforest, one of seven "biomes." The others were: a small "ocean," complete with coral reef and fish; a mangrove wetlands; a savannah grassland; a fog desert (where fog precipitates out and supplies moisture); a marsh system; plus the all-important agricultural area. There was also a human habitat complex consisting of living areas, laboratories, and workshops. Altogether this was a sealed artificial world, on the surface of the planet instead of flying through space, but otherwise entirely cut off from the natural world around it. When complete it was the largest airtight, self-sustaining ecosystem ever built.

# # #

As much of a masterpiece of ecological systems engineering that Biosphere 2 was, the real stars of the show were nevertheless still to come: the "biospherians," as they called themselves (others called them "bionauts"), who would enter the enclosure on September 26, 1991. These were highly motivated, dedicated people, single-minded in their determination to make it all work. They were individualistic by nature, and resilient under pressure, an attribute that they had acquired during a long period of training and testing beforehand. There would be four men and four women. Not all of them were college graduates, nor were all trained scientists. The most credentialed among them was Roy Walford, MD, who would be the team's physician. The others, in no particular order, were: Taber MacCallum, a native New Mexican who would be responsible for monitoring air quality within the enclosure; Jane Poynter, his partner, who would oversee "the Agriculture," as she called it; Mark Van Thillo (nicknamed "Laser"), a Belgian who would be chief engineer on the project; Linda Leigh, a botanist; Abigail Alling ("Gaie"), a biologist who would manage the marine biome; Sally Silverstone ("Sierra"), who would be team captain; and Mark Nelson, who would be in charge of waste recycling and communications.

All of them had undergone a grueling, months-long, training period, first on the research vessel *Heraclitus*, where they learned how to work together in small groups in isolation, plus a year-long stay at Quanbun, a cattle ranch in a remote area of the Australian outback. Jane Poynter regarded the combined experiences as "training in extremes."

Prior to Closure, the official beginning of the Biosphere 2 experiment, Gaie Alling lived for five days, alone, in a small, closed test module. After that, Linda Leigh spent three weeks inside the test module without encountering any significant problems. Then, finally, the grand day arrived.

Closure occurred at 8 o'clock on the morning of September 26, 1991, when, in front of a press crew that had gathered at the site, the eight human specimens, all of them clad in identical blue jumpsuits, walked past them, single file, and approached the airlock outer door. Jane Poynter, who later wrote a book about the project and her participation in it, remembered: "the airlock door slammed shut behind us, and we opened the inner door into Biosphere 2, to silence."

And to problems. Almost from the beginning, very little went exactly as planned. First was the problem of food, of which there was never quite enough. The biospherians ate only porridge for breakfast, after which everyone was very touchy in the mornings. Soon, at every meal, all of the crew members were literally licking their plates clean. They ate so many sweet potatoes that their complexions gradually turned a shade of orange.

"Long before the next meal arrived, I had usually burned up the calories from the last one," Jane Poynter said. "I dragged myself from chore to chore."

There were plenty of chores: the farm had to be weeded every day, and some plant provisions had to be harvested for their next meal. Rice had to be threshed. The chickens, pigs, and goats had to be fed. Morning glories had to be cut down from the spaceframe each day because they blocked light from the rainforest, which reduced oxygen levels and retarded plant growth. They had to monitor and adjust the pH of the ocean by dumping bicarbonate of soda into the water – some 4,500 pounds of it by the end of the program.

In the fall, cooler temperatures caused condensation to build up on the steel skin of the crew habitat, and it began to rain in the hallways and even in their bedrooms. They put buckets down to catch the drips.

Oxygen levels rose and fell with the amount of sunlight received by the plants. On cloudy days, there was less oxygen available, but higher levels of ambient CO_2, which caused headaches. Taber MacCallum analyzed the atmospheric composition every three days or so and discovered that the enclosure was steadily losing oxygen at the rate of 0.23 percent per week. By April of the first year seven tons of O_2 were missing, and they had no idea where it went.

Worst of all from a group interaction standpoint was the gradual descent of the crew into factions. It is well known that small, isolated teams such as scientific crews wintering over in Antarctica often devolved into factions, taking on an "Us versus Them" mentality, and soon enough just such a rift developed among the biospherians. It was never entirely clear what the specific issue dividing the group was. In a paper that Roy Walford wrote afterward, he described the split into factions with "one totally loyal to and supportive of outside management decisions, the other ... increasingly hostile to what they considered an arrogant, scientifically inept, and abusive management team."

Then there emerged a rumor to the effect that the project's Science Advisory Committee intended to replace John Allen, who was the director of research, with someone else. Four of the biospherians were in favor of this: Walford, Linda Leigh, Taber MacCallum, and Jane Poynter, and four against: Gaie Alling, Mark Van Thillo ("Laser"), Mark Nelson, and Sally Silverstone ("Sierra"). By July 1992, 11 months into the first year, it was a case of Us (the first group), and Them (the second). Things were never quite the same after that, but although the two groups managed to get on with their work, they did so within an omnipresent undertone of anger, hatred, and depression.

Jane at times thought she was losing her mind, and eventually she and Taber sought help from a psychotherapist in Phoenix who talked with them by phone, separately, twice a week. The therapist assured Jane that her reactions were understandable given her confined situation, but ever afterward Jane regarded Biosphere 2 as a dystopia, not the mini-paradise she had once imagined it to be.

Oxygen loss continued to the point that some of the crew started taking Diamox, for altitude sickness. On some nights they slept with an oxygen tube strapped under their noses. Finally, in January 1993, in the second year of the mission, the Science Advisory Committee and Mission Control decided to inject oxygen into the enclosure. Over the next two weeks they injected 31,000 pounds of liquid oxygen into the Biosphere. That brought the O_2 level to 19 percent (21 percent being normal in the outside world).

When they could no longer grow sufficient food for themselves, the bionauts began eating some of their excess seed and emergency grain stocks. In a 2019 assessment of the project published in the *New York Times* ("The Lost History of One of the World's Strangest Science Experiments"), science reporter Carl Zimmer said that staff members made regular "deliveries to Biosphere 2, provisioning it with seeds, vitamins, mouse traps and other supplies twice a month." But he did not cite a source for this claim, and while Jane Poynter mentions the occasional smuggling in of beer, wine, rum, and M&Ms, she adds that the amount by which "these few acts of hedonism" increased their overall caloric intake was "wholly insignificant."

On June 2, 1993, their primary freezer broke down. The crew members transferred food items to a backup freezer that Sierra had argued for prior to closure. Without it, they might well have been lost.

Toward the end of the project, the clothing of the crew members had become threadbare; many items had holes, and some of their shoes simply wore out. Many of the crew members had brought in multiple pairs of shoes; still they, too, were wearing out.

With just a month to go, oxygen was again injected into the building.

The end came on September 26, 1993, at 8:20 AM, when the airlock door was opened, and they all walked out, gaunt but otherwise healthy, and wearing the same jumpsuits as the ones they had worn walking in, although the fabric had been taken in to prevent sagging. (Jane Goodall had started a 20-minute speech at 8 AM, adding an additional few minutes to their time inside of what some of them now regarded as "the cage.")

# # #

In summary, over the two-year course of their confinement to Biosphere 2, the crew members underwent constant hunger, increasing oxygen deprivation, divided up into factions, and broke into their emergency food and drink supplies. They were forced to rely on external assistance, principally in the form of oxygen infusions, but also for at least occasional imports of essentially recreational food supplies. They had a mechanical breakdown that would have terminated the mission were it not for the availability of a backup freezer. And all of this in just two years.

Perhaps the most successful feature of Biosphere 2 was the waste and water recycling system, which operated efficiently the whole time. But the nature and operation of such systems was well understood beforehand. The division of the crew into factions was unpleasant for all concerned, but by itself had little effect upon the outcome. As biospherian Mark Nelson wrote in a retrospective article many years later, in 2017:

> The crew continued to work together, and feast and party together, despite such conflict and factions which are well-documented and unsurprising developments in isolated groups of people. That there was no subconscious sabotage of fellow crew members or the overall project speaks to the dedication and skill of the crew, and their understanding, which becomes quite accentuated in a small life system, that Biosphere 2 was their life support system. That is the unexpected and hopeful lesson of the Biosphere 2 crew – not that, like any group or family of humans, that there is conflict and division.

It might be thought that longer the time frame for a mission such as this aboard a starship, the more likely it would be that the same types of problems would occur in more serious forms. But two of the problems that did occur, food scarcity and oxygen depletion, could be addressed and rectified by systems engineering and science. More food could be produced by starting with additional plants and livestock. And there was an explanation for what had caused the progressive oxygen loss: as the soil microbes broke down contaminants, they took oxygen out of the atmosphere and produced carbon dioxide by respiration. Oxygen was also sucked out of the atmosphere by unpainted concrete. But these problems could be solved by better design.

Mechanical breakdowns, by contrast, are a different story. Wear and tear through constant use would only increase with time, occurring more frequently and with increased severity across mission times in the hundreds of years. And as we have seen, reliance on automated repair systems does not solve the breakdown problem because such systems are physical apparatuses that are themselves subject to failure. This is a doom loop that appears to admit of no escape.

Although the factionalism that developed among the Biosphere 2 crew members did not seriously threaten the project's final success, the members were nevertheless eager to escape as the end approached, when, in Jane Poynter's words, all of them were "starving, suffocating, and going quite mad." Certainly none wanted to extend their stay in the enclosure, and when medical consultants from the University of Arizona stated that it would be best if all of the crew remained on the Biosphere 2 campus for three days after their release, supposedly to get them reacclimated to life in the outside world, the crew members adamantly refused to comply. Roy Walford, the physician, argued that this would only add to their stress, as they just wanted to get out of the cage and party. The medical consultants withdrew their request.

But the success of the project after two years gives us little basis for extrapolating out to significantly longer time periods such as would be sustained by a world ship headed across light years of space. The experience of Biosphere 2 is also silent in regard to the potential success or failure of a much larger crew population: what happened to a crew of eight tells us little about the fate of a far larger crew of, say, 800. The fate of the biospherian crew had been to survive all of their hardships. But the likely fate of a bigger and longer trip is a much dimmer and less optimistic prospect.

Taking a commonsense view of the matter, it does not seem that there is a magic number or golden-mean crew size in which conflict is absent or minimized. Factions can and do occur on any and all scales and population sizes, whether it be among the 535 members of the US Congress, or a much larger group, such as a nation as a whole. Civil wars, for example, have existed throughout history, and they are among the bloodiest and bitterly contested wars ever fought. Human beings are inherently conflict-prone on Earth, and are not likely to be less so within the confines of a manufactured biosphere traveling for decades through the blackness of space. And so in the end it is difficult to regard with assurance the prospects of a multigenerational interstellar mission consisting of hundreds of people and lasting for hundreds of years.

9 THE MORAL STATUS OF THE TRIP

A multigenerational interstellar voyage divides neatly into three distinguishable sets of populations, each of which faces its own particular challenges (Table 9.1). First is the group that boards the spacecraft and exits the solar system forever, the Departing Generation, or Population 1 (Gen a). This is the echelon whose members will experience the loss of the natural world and all of its pleasures and pains, freedoms and restraints. Following them in time is the set of generations that is then born on the spacecraft: an initial generation (Gen b), plus their series of descendants, including those who make up the penultimate generation (Gen n), the last but one to be born aboard the spacecraft prior to its arrival at the extrasolar planet. These make up an unknown number of transiting generations. Their sole function is to act as caretakers of the ship, keeping it intact and operating efficiently until that responsibility can be handed off to the next generation in sequence, and so on down the line. This is the Caretaker Generation, or Population 2. Finally, there is the Arriving Generation, Population 3 (Gen x). Its members will be the only group of individuals to set foot upon the destination planet.

Population 1 would be the sole group whose members would have had personal experiences and memories of the natural world. It is likely, given the known mental health issues of astronauts in Earth orbit, that during their trip toward the stars, members of this first generation would suffer considerable emotional stresses and strains in the form of pangs and longings for the many remembered comforts of Earth, now lost

Table 9.1 Sets of populations aboard a multigenerational interstellar spacecraft.

Population 1 Departing Generation (Gen a)	Population 2 Caretaker Generations (Gen b, c, d, ... Gen n)	Population 3 Arriving Generation (Gen x)

to them forevermore. Further, a distinguishing characteristic of Population 1 is that it is the only group whose members would have *volunteered* for the trip, who made a free and informed decision to board the spacecraft where they would live out the remaining years of their lives.

But once that group is fully replaced by the first generation of Population 2, Gen b, any lingering emotional tolls and losses experienced by members of the previous generation, native to Earth, would have vanished. Members of Gen b, and all the rest to follow in sequence, would have no nostalgia or homesickness for life back on Earth, because they never had any firsthand experience of Earth to begin with. They may of course experience stresses of their own, consequent upon their being confined to the spacecraft for their entire lives. A further characteristic of Population 2 is that none of its members *chose* their fate: they are there aboard the spacecraft involuntarily, whether they like it or not. It is also true that all the members of these multiple, successive generations would live out their lives for a cause that they did not voluntarily accept, and, in the end, would die aboard the vehicle without ever seeing the voyage completed.

An additional feature of the successive generations of Population 2 is that its members may become progressively less fit as new generations come into being. Even if we assume that there will be minimal radiation damage to their genomes, nevertheless such factors as random mutations and genetic drift make it probable that there would be some declines in the health and fitness among their progeny, and perhaps also a deterioration in their mental capacity, physical skills, and general intelligence. Genome mutations would cause disease, immune systems would decline with age, cancers and tumors would grow, and viral and bacterial organisms would evolve and have their ill effects aboard the spacecraft. These impairments might make each consecutive generation somewhat less mentally and physically equipped than the one before it in this closed, contained, and radically isolated society of spacefarers.

There would probably be some linguistic and cultural changes across the generations as well. In her 2003 paper, "Language Change and Cultural Continuity on Multi-Generational Spaceships," Sarah G. Thompson, professor of linguistics at the University of Michigan, suggests that certain common terms would fall out of use. "Words like snow, windy, river, ocean, mountain, sunburn, summer, winter, horse, tiger, and ostrich are likely to be non-useful in the travelers' world, as well as words for non-portable Earth-bound cultural artifacts like car, train, boat, truck, airplane, skyscraper, tunnel, bridge, and so forth." The spacefarers would also be somewhat limited in their activity options. "You can't travel to new and different places inside the spacecraft, or meet any people other than the on-board population." There would be a distinct lack of new and interesting restaurants.

The far-off generation that constitutes Population 3, the Arriving Generation, would be the only group to set foot on the destination planet. There are at least two possibilities as to how members of that generation would regard that event. One, some of them might dread the prospect. After all, these people and many of their ancestors have spent their whole lives aboard the spacecraft. They are accustomed to it, and it is the only world they know – or may want to know. Venturing upon an alien planet might be a frightening prospect, similar to the old mariners' fear of sailing beyond the horizon. It is more likely, however, that the travelers would have been sufficiently educated about the upcoming experience and would welcome the opportunity to leave the ship, alight on the new planet, and make their homes there.

However, even if they arrive safely, there is no assurance that the founding population would survive and prosper. Startup colonies often fail, as did the Roanoke Colony in North Carolina, which simply disappeared, and the Jamestown Colony in Virginia, which was abandoned for a time by its inhabitants after 80 to 90 percent of them died of starvation and disease. A similar fate could befall the arriving population of an extrasolar planet. The physical constraints of space travel, after all, are far more stringent and instantly lethal than at any location that humans have attempted to colonize on Earth. The inhabitants of Earthly colonies at least breathed the same air and lived under the same sun as in the locations they departed from. Colonists of an extrasolar planet, by contrast, would be faced with an entirely foreign and unfamiliar environment, and would be in for some unpleasant shocks and surprises that they might or might not be able to manage.

# # #

A vivid picture of what such a voyage might be like, and how it might end, is offered by the science fiction writer Kim Stanley Robinson in his novel *Aurora* (2015), which presents a scenario that is about as likely as any other.

The starship involved holds a population of about 2,000 people, plus all the life forms needed to sustain them, parceled out in 24 different biomes. The ship is powered by a nuclear fusion reactor and travels at 10 percent the speed of light. Artificial gravity is provided by means of rotation of the crewed toruses, of which there are two, linked together by a long cylinder, "like two wheels and their axle." The ship is not a very pleasant place to be: there are surveillance cameras everywhere, there is little privacy, and everyone's movements are carefully tracked. It is a de facto dictatorship, and not all aboard are happy to be there. "This place is a prison," one of them says.

The ship reaches its destination during the seventh generation, which scores lower on intelligence tests than did the previous generations. On reaching the target planet, called

Aurora, a member of the Tau Ceti system, most people were eager to leave the ship: "They wanted down." But not all of them. "Some confessed they were afraid. ... Who needed bare rock, on a lifeless moon, on the shore of an empty sea, when they already had this world they had lived in all their lives?"

Those who landed encountered a few surprises: they were bothered by high winds on the new planet. Also, the place was overrun with bugs, as well as by new and deadly pathogens (prions) to which the settlers had no immunity.

A week after landing, more than half the people had fevers. And then some started to die, "of something like anaphylactic shock." This was followed by a rash of suicides.

As Prussian Field Marshall Helmuth von Moltke is reported to have said, "No plan of battle survives first contact with the enemy." And all at once this new exoplanet had become the enemy. Some residents argued for getting away from it, by resupplying the ship and returning to Earth. Others wanted to reboard the ship and live on it in perpetuity.

The suicides continued. The remaining inhabitants took a vote, revealing that the members had predictably split into two factions. The "Back to Earth" option won by a narrow margin. So half of the population remained behind on the planet, in an attempt to terraform it, while the others left on the newly outfitted spacecraft.

But vital supplies became low during the course of their return. The cows, which had been reduced to the size of dogs by genetic engineering, were slaughtered and eaten, and it was soon obvious that a "general famine was causing serious malnutrition in the human passengers of the ship."

They are saved by a deus ex machina plot twist in the form of information from Earth that Russian scientists had perfected the technology of suspended animation. The crew members now themselves develop and utilize it in order survive until they reach the solar system. The sleeping space travelers are awakened in time for a descent to Earth, led by a woman named Freya.

After having settled themselves on Earth, Freya and some others are invited to a conference about further attempts at interstellar migration. Everyone there seems to be in favor of making additional flights to the stars. "'We are going to keep trying,'" the group's moderator says. "'It's an evolutionary urge, a biological imperative.'"

But Freya, who is a strong opponent of further missions to the stars, is disgusted by this, rises from her chair, and physically assaults the speaker before being led away and out of the room.

At the time of the book's publication, author Kim Stanley Robinson gave an interview to Space.com in which he explained his motivation for writing *Aurora*.

[I wanted to] make the case that although the solar system is in our neighborhood, so to speak, and we can definitely visit it and set up scientific stations all over the solar system, that going to the stars – the new data about what we are as bodies (our bodies are biomes) – made me begin to question the starship project, begin to worry that the stars are simply too far away, even the closest ones.

And he became progressively more convinced that there's actually no intelligible or rational reason to make the attempt. "I think there's a certain craziness to it, or pointlessness – the point seems to be religious, and having to do with species immortality or something like that."

The story told in *Aurora* is a tragedy, unpleasant to read, and would be even more unpleasant to live through in real life. The sad part about it is that much of it could actually happen.

# # #

While the fate of a multigenerational interstellar population cannot be predicted with anything approaching certainty, the many dangers presented by the instantaneously lethal environment of space, plus the interpersonal pressures and conflicts that might result in social breakdown, make it doubtful that a successful transit to another star system with all the successive onboard generations remaining safe, healthy, and happy across time, is a realistic possibility. It is far more likely that the crew would suffer one or another kind of irremediable catastrophe en route than that everyone aboard would survive, and that the final, arriving generation would get there intact. But if that is true, then the question arises whether it would be morally justifiable to launch such an expedition to begin with, given its immense costs, high probability of failure, and lack of any benefit accruing to the sponsors back on Earth who had paid for it all.

In contrast to the issues already considered, all of which are addressed and potentially answerable by science, the question whether it would be morally acceptable to launch a multigenerational interstellar expedition is not a scientific question and cannot be answered scientifically. Science deals with matters of empirical fact, and its disputes are resolved by means of experimental tests. Moral questions, on the other hand, are philosophical in nature, and a fundamental limitation of philosophy is that there is no objective decision procedure by which its disagreements can be resolved conclusively one way or the other. Philosophy proceeds by means of rational argumentation and logic, but the problem is that for every argument leading to one conclusion there is a relatively plausible counterargument leading to another (and often an opposite) conclusion. This does not mean that we should avoid the point at issue or pretend that it doesn't exist, for when a serious moral problem arises it is best to tackle it forthrightly and attempt to provide an answer even if that answer cannot be proven beyond all doubt. Perhaps, as in a court of law, we may accept an answer when it is established beyond a *reasonable* doubt.

Philosophical argumentation also often resorts to reasoning by analogy, by comparing two different sets of conditions and showing that if one of them has a certain moral status, then so does any relevantly similar situation. By definition, analogous cases are never identical, and so a comparison between the two is never watertight. Nevertheless, the closer the degree of similarity between the two cases, the stronger is the conclusion based upon the analogy.

The primary moral question posed by the prospect of a multigenerational interstellar expedition is whether it is morally legitimate to confine a series of generations to a closed and largely synthetic environment for their entire lives without any of the individuals (except for the members of the Departing Generation) having given their free and informed consent to such confinement. This specific issue pertains only to Population 2, the Caretaker Generation, for all of these people would have been born on, and would be forced to live out their entire lives on a space vehicle, in an artificial environment, to which they would be restricted without having agreed to those conditions at any point.

At first glance, and on a completely intuitive level, it seems obvious that such treatment would not be ethically acceptable, for this is a situation that would be morally equivalent to locking up large numbers of people inside a closed and artificial ecological setting and throwing away the key. Such a practice appears to violate their basic human rights in a fundamental way. Doing so would be a case of interstellar kidnapping, subjecting generations of space travelers to involuntary exile for the whole of their lives. If, analogously, the biospherians were suddenly told that they could never leave their habitat, that they must live the remainder of their lives inside the enclosure, and die there, the inhabitants would have an excellent case that this is a form of involuntary imprisonment, and is flatly illegal and morally wrong. And if that would be true of locking up the biospherians, why wouldn't it be equally true of all the members of Population 2 aboard a hypothetical starship traveling through space?

Appearances aside, there is a difference between the two cases that makes them non-analogous. And this is that the biospherians entered their closed environment under the explicit condition that the experiment would last for two years, and then end, after which they would emerge from confinement and return into the world at large. If that condition were suddenly abrogated without their consent, it is indeed true that their rights would be violated. But in the case of an interstellar spacecraft, the members of Population 2 did not have any such prior condition in place, which means that the two situations are different in principle.

The situation of the Population 2 and 3 crew members is in fact more closely analogous to the birth circumstances of every person ever born on Earth, for, without exception, no one on Earth voluntarily chose the living conditions into which they were born. Indeed, in this most fundamental of ways, planet Earth and the multigenerational spacecraft are reasonably comparable: nobody in either case chose the conditions into which they were born; in each case it was a matter of chance.

Still, it could be argued in opposition to this that humans have a natural right to be on Earth, since that is where they were born, and that confining anyone to a lifetime aboard a spacecraft violates their right to be on Earth and is therefore immoral. But if being born on Earth gives you a right to be there, then why doesn't being born on a spacecraft give you a natural right to be *there*?

A second counterargument to the view that being born on Earth entitles the newborn child to remain there is that a spacecraft is no place for any human being to spend his or her entire life because the craft's environment is constricting, artificial, and sterile. But so are many environments on Earth even today: many large, densely populated cities the world over offer constricting, barren, squalid environments, particularly among their poorest inhabitants living in profoundly disadvantaged and impoverished circumstances. Children are routinely born into communities lacking the basic underpinnings of life such as clean water, proper sanitation, and adequate levels of nutrition. While in theory they might be free to leave those communities, they are nevertheless confined in place de facto due to their poverty, lack of access to transportation, and with no place else to go.

Further, we do not normally regard procreating in some constricting environments to be immoral. For example, when it was still a partially walled city, the residents of East Berlin had the right to procreate there even if it was likely that their children would have to remain there for their lifetimes. For that matter, it is not entirely clear that bearing children even in conditions of extreme deprivation is morally wrong.

Childbearing occurred even in Russian forced labor camps; and slaves in the United States (and in other countries) often bore children prolifically, even if there was little hope of their ever being set free.

Still, an additional argument that procreation aboard a spacecraft would be morally wrong is that there is no possibility of escape, that the spacecraft is in effect, and in reality, "a prison." In "An Essay on Extraterrestrial Liberty" (2008), the British astrobiologist Charles S. Cockell remarks that "It would not be inaccurate to say simply that children born in space will be the first humans to be born in cages." (In fact it *would* be inaccurate: in the United States alone, in each year more than a thousand pregnant women give birth while incarcerated.) But until the advent of spaceflight, there was no possibility of escape from Earth, either. There is an obvious and substantial disanalogy between the two cases, however, consisting of the enormous differences in size, richness, and diversity of the two respective environments, for the Earth is vast and open whereas the spacecraft is minuscule by comparison, and closed. But even though the spacecraft is small as compared to Earth, from a moral standpoint a lot depends upon what conditions aboard the spacecraft would be like.

An analogy that has been suggested by some interstellar travel advocates is that of life aboard a large and luxurious cruise ship. The Royal Caribbean cruise liner *Wonder of the Seas* has a capacity of 6,988 passengers. With a crew of 2,300, the total shipboard population is 9,288, all of them squeezed onto a vessel that is 1,188 feet long, 211 feet wide, and with a total of 18 decks. Even though the ship is small compared to proposed multigenerational spacecraft whose lengths are measured in miles rather than feet, no moral issues are raised by the cruise ship voyage for the simple reason that all aboard are in principle and in practice free to leave the ship, and the cruise itself, at any time. That is not true of the members of Population 2 aboard a multigenerational spacecraft, who cannot escape either in practice or in principle. They are just stuck there forever.

Counterintuitively, however, this does not mean that the involuntary confinement of the spacecraft's crew members violates a basic human right to be free. Everything hinges on the physical and social conditions aboard the spacecraft, which proponents of interstellar travel describe as not even remotely as constricting as those into which many people throughout history have been born on Earth. Poverty, scarcity, and deprivation have been the lot of most of the children who have ever been born on Earth, particularly in places, and during time periods, that have been rife with disease, famine, and early death.

In their book *Emigrating Beyond Earth: Human Adaptation and Space Colonization* (2012), authors Cameron Smith and Evan Davies argue that "Regarding the material conditions into which children are born, it is immediately evident that the entire set of material conditions of modern, Western civilization are not necessary for a happy existence; many human cultures over thousands of years and across the globe have had rich and fulfilling lives without, for example, three cars per family."

And in fact, spacecraft crews are routinely described by interstellar travel proponents as being well provided for, entertained, and as being paragons of good health throughout their lifetimes, something that could not be said of most of the children who have ever been born on Earth.

It is difficult to answer the moral status question in any absolute sense given the fact that we do not know the size of the population or the degree of material comfort that would be enjoyed by the spacecraft crew members. We do not know the range of their available opportunities, or the level of reproductive freedom that would exist aboard the starship. We do not know the challenges the crews would face across the many years of the journey, nor their fate upon arrival.

In his paper "Worldship Ethics 101: The Shipborn" (2018), philosopher James Schwartz argues, concerning crew size, that "The population ought to be large enough to be stable in a genetic sense. But ideally it should be large enough to afford the shipborn with opportunities for meaningful intellectual growth, freedom of education, freedom of vocation, freedom of gender identity, freedom of sexual orientation, and reproductive autonomy." This last is to say that women must be allowed to have children aboard the ship, but must not be compelled to do so in order to satisfy a given ideal population requirement. But what this ideal size is, and how it is to be realized in a closed spacecraft, the author does not say.

In general, the moral status of a multigenerational voyage is contingent upon and proportionate to the level of potential harm or degree of disadvantage experienced by the crew members. The better the life that can be provided to them, the more morally permissible is the voyage. There is a certain threshold of adequate living conditions, comfort, and personal autonomy below which an interstellar voyage would not be morally permissible, but exactly what that threshold is, is difficult to specify in advance.

Other things being equal, the larger and more well-appointed the ship, the more ethically justifiable the voyage becomes. But the bigger the population, the more likely it is that it would harbor members with destructive and violent impulses.

It is also true that the larger, more populated, and more luxurious the ship, the greater are the resources needed to develop it. And if there is no urgency about making the voyage, then why make the trip and confine all the successive populations that would be needed to complete it? A very large and expensive spacecraft then becomes economically suspect, as a plain waste of money, time, and resources that could be better spent on more immediate and pressing needs the realization of which would yield greater near-term benefits. This in turn raises the question of cost: even if the voyage is morally justifiable, and we have as yet to see any conclusive evidence that it isn't, is it worth what it would cost to plan and build the ship, and launch it toward its extrasolar destination?

# # #

One of the least-discussed aspects of interstellar flight among the ranks of its proponents is the overall cost of the enterprise. This is not surprising. After all, we are talking about a spacecraft whose size and onboard population we do not yet know, powered by a propulsion system yet to be invented, headed to an extrasolar planet that has yet to be identified, which is therefore at an unknown distance away from us, traveling at a velocity that is an unspecified fraction of the speed of light, and for an undetermined period of time. Since many if not most of the postulated technologies involved are entirely hypothetical and imaginary, and would be deployed on an engineering scale never before attempted, there is no reliable way by which their costs can be rationally assessed. Doing so would be like pricing out a dream.

And even when some of the relevant factors *are* provided, no matter how hypothetical they might be, such cost estimates that are offered in interstellar vehicle proposals rest on improbable assumptions or appear to be pulled out of thin air. What is more, some of the estimates are so very large in magnitude that the scientists advancing them fall back on the use of exponential notation (as for example, 10^{15}) instead of whole numbers. This is because if they were expressed as whole numbers they would be made of so many zeroes in sequence that they would be unintelligible to the average reader. They are figures that just numb the senses and paralyze the brain. Quantities expressed in exponential notation also impart a spurious impression of scientific accuracy and authenticity to the figures.

Alternatively, some cost estimates are given as multiples of GNP, or gross national product. And so for example in 1968, when Freeman Dyson attempted to estimate the cost of an Orion-type H-bomb-propelled spacecraft that "could reach many nearby stars in the course of a few centuries," he said: "If we continue our 4% growth rate we will have a GNP a thousand times its present size in about 200 years. When the GNP is multiplied by 1,000, the building of a ship for 10^{11} will seem like building a ship for 10^8 today."

In this case, the ship Dyson is talking about is his "conservatively designed space ship" that is powered by, as he puts it, "3×10^7 bombs," that is, 30 million bombs. The problem with this cost estimate is that neither Dyson nor anyone else can reliably predict what those bombs would cost at the time the ship would be built, 200 years from now. Indeed, it is difficult enough to predict the price of a barrel of crude oil in a year's time, much less what that cost would be in 200 years. Prices are affected by innumerable factors including supply and demand, inflation, deflation, recessions and depressions, wars, global epidemics, tariffs, taxes, government regulations, strikes, shortages, transportation costs, resource depletion, and so on. For Dyson to accurately predict the cost of a starship 200 years from now is the mystic equivalent of the Wright brothers, in 1903, predicting the cost of a Boeing 747 in 1969, the year of its first flight.

Later, in 1984, when Anthony Martin wrote in the *Journal of the British Interplanetary Society* about the cost of interstellar vehicles, he stated: "Based on the technology of today Project Daedalus may cost about 10^{13} in very round figures," and then spoke of advancing "a crude guess" as to when it may be possible to afford interstellar flight. The Daedalus rocket was to be powered by a controlled nuclear fusion propulsion system. Such a system does not as yet exist, and we simply do not know its cost, nor whether, nor when, it could be built. Its cost in the indefinite future cannot possibly be pinned down today.

Later still, in their authoritative review, "World Ships: Feasibility and Rationale" (2020), a group of four authors stated as their summary of world ship costs, that "building and launching a world ship would require two economic conditions to be satisfied. First, a solar system-wide economy with large-scale in-space manufacturing capabilities. Second, GDP growth rates of 2%/year or higher need to be sustained for the next 500 to 1000 years."

If a solar system-wide economy is in fact needed to build and launch a world ship, then we are further away from realizing the dream of interstellar flight than ever before imagined. Just how far away can be judged from the fact that, as we have seen, since the end of the Apollo program with the Apollo 17 flight in 1972, no human being has again set foot on the Moon, much less stepped out upon the surface of another planet. That is to say, in the more than 50 years since the last moon mission, we have made only minuscule progress toward occupying and industrializing the solar system. At that rate, the goal of a fully developed solar system with a large-scale space manufacturing system in place all over it, could well be a thousand years away, if indeed it is ever achieved at all.

There is an additional problem with existing cost estimates for interstellar missions, and this is that as approximate and imaginary as these estimates are, they are almost certainly *underestimates*, as has been true of most of the cost estimates of megaprojects ever undertaken in the past. And a multigenerational space vehicle would the most gigantic megaproject in all of human history.

Assessing the costs of megaprojects falls within the scope of project management economics. A 2014 paper, "What You Should Know About Megaprojects and Why," by Bent Flyvbjerg, professor and founding chair of Major Program Management at Saïd Business School, Oxford University, offers several important insights about how it happens that megaproject cost estimates so easily get out of control.

First, he describes what, in general, megaprojects are, to wit: "Megaprojects are large-scale, complex ventures that typically cost US\$1 billion or more, take many years to develop and build, involve multiple public and private stakeholders, are transformational, and impact millions of people. ... Megaprojects are a completely different breed of project in terms of their level of aspiration, lead times, complexity, and stakeholder involvement."

It is characteristic of megaprojects that their appeal is a product of *emotion*, in feelings of the sublime, in "the rapture engineers and technologists get from building large and innovative projects, with their rich opportunities for pushing the boundaries for what technology can do, such as building the tallest building, the longest bridge, the fastest aircraft, the largest wind turbine, or the first of *anything*."

Their being largely a product of emotion rather than rational investigation then leads to problems in implementation. Megaprojects are inherently risky due to their long planning horizons and innate complexity. That complexity in turn leads to delays in execution and to cost overruns.

The author provided a list of 33 megaprojects and the amount by which their actual costs exceeded their projected costs. The Suez Canal in Egypt, for example, had a cost overrun of 1,900 percent; the Scottish Parliament Building, 1,600 percent; and the Sydney Opera House, 1,400 percent, and so on down the list.

But there is a fundamental disanalogy between past cost estimates for megaprojects and cost estimates for something like a starship. Whereas many of the off-the-shelf component parts and building blocks of projects such as the Sydney Opera House and the Scottish Parliament Building have been used throughout history, the same is not true of starships, whose propulsion system, physical structure, the makeup of its component parts, and so on, are all unknowns. If the system being priced is essentially conjectural, then so too are its projected costs.

We may illustrate how routinely rockets and space-related systems encounter delays and cost overruns by citing NASA's experience with the Space Launch System (SLS) and its Artemis rockets, currently under development. The SLS program was designed to establish a human presence on the Moon and, ultimately, also on Mars. The first launch of the Space Launch System was originally set for 2016, but was rescheduled and not finally launched until November 2022, as an uncrewed Artemis I rocket. The program was also having trouble keeping track of costs, and in September 2023, the US Government Accountability Office undertook a study of the program's progress to date. In its report, "Space Launch System: Cost Transparency Needed to Monitor Program Affordability," the government stated that, "While Artemis I was ultimately a successful launch, the SLS program has faced a variety of challenges, which led to significant cost growth and years of delays to that launch .... Senior NASA officials told GAO that at current cost levels, the SLS program is unaffordable," and that "at current cost levels the SLS program is unsustainable."

But if this system is unsustainable, then in all probability, so would a giga-scale megaproject such as the construction, outfitting, and launch of a multigenerational interstellar spacecraft.

10 LET US HIBERNATE

It's easy to become somewhat disillusioned about the prospects of interstellar travel in view of so amazing a quagmire and botheration that surrounds the technical and practical feasibility of a multigenerational ship, the iffy moral status of embarking upon such a voyage, the truly otherworldly costs of getting the whole project up and running and, finally, the likelihood of sending it off to the suitably earthlike extrasolar planet of choice. We can be inordinately grateful, then, that there's a possible solution to these problems that could resolve many of them very neatly in one fell swoop. And that is simply by putting the crew into a state of suspended animation and letting them sleep their way to the stars.

Indeed, if there are two words that pop into the mind at the mere thought of star travel, they are "suspended" and "animation." Everyone's heard them, and everyone knows what they mean, which is: putting people to sleep – or at any rate into a sort of extra-hyper-super-duper deep sleep, deeper by far than mere ordinary nighttime slumber.

Suspended animation has a number of key advantages over a multigenerational flight in which the entire onboard population is alive, awake, and functional, consuming food, creating waste products, and so on. A "sleeper ship," in which all or most of the population is in a state of suspended animation is not encumbered by any of that. Instead, since there's no longer any need for establishing and maintaining numerous biomes, animals, plants, feedstocks, and other provisions to keep several generations of people alive and well, the entire trip could be made much more cheaply and with a smaller and simpler spacecraft than with the multigenerational, live-action version. Also, putting people to sleep avoids any possibility of crew factionalism, social breakdown, the potential threat of crew mutiny, and other such embarrassments. It escapes the problem of boredom and depression that is otherwise associated with very long-term stays inside closed habitats. Further, there is evidence that people in a state of hibernation are less susceptible to radiation damage to the body. Altogether, this looks like a winning proposition.

Of course, suspended animation has its own set of problems, but in the esoteric realm of interstellar travel theory and discourse, what doesn't? One difficulty is that despite the reduced size and mass of the spacecraft, a vehicle with a crew in suspended animation is still a gigaproject and therefore would entail all the uncertainties, delays, unforeseen difficulties, and cost overruns that are characteristic of exceptionally large-scale undertakings of any type. There's also the issue of dealing with a potentially catastrophic emergency onboard a ship full of people all of whom are blissfully and peacefully asleep.

Suspended animation, or something like it, has a long history in both science fiction and in popular culture. It appears in fairy tales such as *Sleeping Beauty*, which has been incarnated as a ballet, an opera, a Disney cartoon, television shows, musicals, poems, and in video games, and has also been depicted in many works of art. Then there is the classic short story "Rip Van Winkle" by Washington Irving in which a British man traveling in the American colonies drinks some native moonshine and falls into a 20-year-long deep sleep. When he awakes he finds that society has changed substantially in the interim. Suspended animation is of course a staple of science fiction, depicted in movies from Woody Allen's comedy *Sleeper*, to Stanley Kubrick's *2001: A Space Odyssey*, to Christopher Nolan's epic *Interstellar*, and in many other space operas. In none of these cases has the sleep in question been the kind of normal and natural sleep that we fall into every night; rather they are long-term, unnatural states that have been induced by deliberate intervention, whether by magic, as in *Sleeping Beauty*, or by cryopreservation, as in *Interstellar*, or by other means.

In medical science, suspended animation does not denote a single, unitary concept but rather represents a continuum of states of consciousness in which an organism's activities, metabolic processes, and body temperature are all reduced to one extent or another. These range from, at one extreme, chemically induced deep sleep, to an array of intermediate states of hibernation, to, at the other extreme, cryonic freeze-preservation. The idea of cooling the human body as a therapy for traumatic injury is an old one, and goes back to the ancient Greek physician Hippocrates who advocated packing wounded soldiers in snow and ice for transportation to army hospitals. In 1810, Napoleon's military surgeon, Dominique Jean Larrey, observed that injured soldiers who were kept close to a fire fared less well than those who were left in cooler conditions outside of their encampments.

The least effective form of suspended animation, induced semi-natural sleep, would not be suitable for purposes of space travel in any case. The reason is that all the basic metabolic activities would continue on during induced natural sleep, even though at a slower pace. Over the course of prolonged semi-natural sleep, men would grow exceptionally long beards, and everyone's hair would grow unstoppably. People's fingernails and toenails would also propagate to unpleasant lengths, as indeed happened to Howard Hughes, the aviation pioneer and billionaire who late in life lived as a recluse, confining himself to a hotel in Las Vegas and letting his beard, hair, and nails grow to absurd lengths.

Further, protracted induced sleep comes with many other unpleasant consequences: nasal mucus would accumulate and would probably block airflow if not removed. Digestion would continue, albeit more slowly than usual, and waste matter would also accrue and would require removal. Due to lack of exercise, the joints would eventually harden and freeze up. Muscles would atrophy, and bone matter would demineralize. Due to the lack of a varied sensory input, eardrums would ossify, and the vestibular system would essentially freeze in place. From lack of use, the retinas might progressively deteriorate.

At the other extreme, cryonic preservation is made superficially plausible by the fact that some animals actually freeze solid during the winter months and then spontaneously revive in the spring. There are many insect species, as well as certain species of amphibians, mainly frogs, and also reptiles that are freeze-tolerant. While in the frozen state these animals stop breathing and their hearts stop beating. They can endure the conversion of 50 percent or more of their total body water into extracellular ice and then employ a suite of adaptations that counter the negative effects of freezing. In addition, their bodies contain natural cryoprotectants, a type of biological antifreeze, that minimizes cell shrinkage and resists cell membrane deformation. The net result of such modalities is that these creatures can reactivate all of their vital functions and resume normal life after days or weeks of continuous freezing.

Of course, humans do not possess such abilities, but that doesn't mean they cannot be frozen, at least as embryos, and then when thawed can resume normal growth and development, and can be successfully implanted into the uterus. Also, frozen sperm cells have spent years of existence in liquid nitrogen, at the temperature of minus 196 °C, and are then revived after thawing.

There are also adult humans who have been placed in liquid nitrogen at –196 °C and exist in a state of cryonic suspension. But these people – "patients in biostasis," in the lingo of the cryonics business – were already dead to begin with, and it is entirely conjectural (and indeed quite doubtful) whether they could ever be reanimated and their damaged cells be repaired by the as-yet-to-be-invented, hypothetical science of molecular nanotechnology (atomically precise manufacturing). Cryonics is a borderline fringe idea, more fantasy than science, and offers scant hope to interstellar travel contenders, or for that matter, anyone else.

On the other hand, there are several accounts of "miracle" survivals after individuals have endured long periods of semi-frozen, or at least super-cooled, existence. There was the case of Mitsutaka Uchikoshi, who after a 2006 hiking accident in Japan was found unconscious 24 days following the mishap. His body temperature had dropped to 22 °C (71 °F), and he had no detectable pulse. But he was brought to a hospital and made a full recovery. The doctors theorized that he had fallen into a state of hibernation. There was also the remarkable case of a 16-year-old boy who in 2014 stowed away in the landing gear wheel well of a Boeing 747 en route from San Jose, California to Hawaii. The plane's cruising altitude was 38,000 feet, and the outside air temperature dropped to a low of –85 °F. After spending about five hours in very cold and oxygen-deprived conditions, he revived after landing and recovered with no medical complications. Attending physicians speculated that the boy's body had entered a state similar to hibernation which allowed him to survive extreme cold and minimal oxygen levels.

Suspended animation is such a diffuse and multivariate concept that, at a stretch, it even includes states such as stupors and comas, conditions in which people are unconscious to a degree that they do not respond to external stimuli and cannot be awakened. Still, there have been recoveries from even long periods of being comatose. A man named Terry Wallis existed in a minimally conscious state (MCS) for 19 years before spontaneously awakening in 2003, in a rehabilitation center in Mountain View, California, and uttering his first word, which was "Mom." Brain scans taken afterward showed that Wallis's brain had apparently grown a mass of new axons, nerve fibers that conduct nerve impulses between neurons.

Whatever the possible application of suspended animation might be to space travel, one does not want to induce conditions such as minimally conscious states, or, worse, persistent vegetive states, among crew members, either deliberately or accidentally.

# # #

Hibernation per se, however, is a different story, for it is a separate category of suspended animation that is relatively rare in the animal kingdom, but has been successfully induced in humans in specialized medical settings. The state of hibernation is characterized by reduced activity level and metabolic rate, and also by a reduction in body temperature. It is well known that black bears hibernate during the winter months, as also do ground squirrels, marmots, prairie dogs, groundhogs, and dwarf lemurs, which are found only in Madagascar. Further, a few species of birds are thought to hibernate. In nature, hibernation functions as an energy-conserving measure among animals when sufficient food is not available. (Still, bears occasionally die while hibernating.)

These various states of lowered body activity, and reduced metabolism and temperature have led scientists to wonder if hibernation, or something like it, could be used in spaceflight, as on a trip to Mars. During the 1990s medical scientists had recourse to an old term, "torpor," to describe the condition of being in a state of lowered physiological activity characterized by reduced metabolism, heart rate, respiration, and body heat. Scientific interest in torpor stemmed in part from the increasing number of people recovering from experiences during which their bodies were cooled in such a way that saved their lives: the case of Mitsutaka Uchikoshi, described above, for example. There was also increased use of low-temperature surgery, which is now a state-of-the art medical procedure involving the use of therapeutic hypothermia, a treatment that lowers a patient's body temperature to reduce the risk of ischemic injury (restricted blood flow) to tissues.

Patients who have suffered traumatic injuries regularly undergo therapeutic hypothermia for periods up to one or two days, during which time they are in torpor; afterward, they are returned to normal body temperature, and then go on to make full recoveries. The increased successful use of this methodology in critical care medicine has led space scientists to ask if this short-term medical practice might be extended for much longer periods. Currently, the longest time during which a patient has been in a state of therapeutic hypothermia is 14 days. The hope was that perhaps progress in medical understanding and use of this procedure could allow it to be sustained for weeks or months at a time.

In 2013, NASA itself showed interest in the use of hibernation when it issued a grant to an outside contractor, SpaceWorks Enterprises, of Atlanta, Georgia, to research the possibility of using therapeutic hypothermia to induce and maintain torpor among the crew members of Mars missions. Head of SpaceWorks was John Bradford, a PhD aerospace engineer who had led several NASA and DARPA projects designing military spaceplanes. Bradford was as committed as Elon Musk and other visionaries to the idea that we must establish a human presence on Mars. "We're not in the vein of an Apollo mission anymore," he said. "No more 'flags and footprints.' We need to become a two-planet species."

By April 2014, Bradford, together with Doug Talk, an obstetrician who had used therapeutic hypothermia to treat oxygen-deprived infants, and Mark Schaffer, an aerospace engineer, had written and submitted to NASA their report, "Torpor Inducing Transfer Habitat for Human Stasis to Mars." This was a 73-page-long document in which the authors discussed a range of medically accepted protocols for putting humans into a state of therapeutic hypothermia, or torpor, as well as a means by which adequate nutrition could be supplied to the crew members across the full duration of the hypothermic state. The authors wrote:

> All the nutrition and hydration needs for a person can be provided by a liquid solution and administered through an intravenous (IV) line directly into the body. This solution is known as "Total Parenteral Nutrition" or TPN. This aqueous solution contains all nutrients that the body needs to maintain full physiologic function. The solution is fed slowly through a permanent IV line to the body.

A major advantage of the TPN system is that since all the nutrients are in liquid form, there is no solid waste matter that needs to be periodically removed; all of the waste products are liquid and can be drained by catheter. The authors noted that, in some extreme cases, patients have had all of their nutritional needs met by TPN for periods lasting more than a year. They therefore proposed the use of the TPN system during a crewed Mars mission. In addition, they designed a crew support system that would hold the crew members in place, supply nutrients, and carry away liquid wastes. A thermal management system would be used to cool the body by means of a small plastic tube that would be inserted into the crew member's nasal cavity and deliver a spray of coolant mist. The mist would evaporate directly underneath the brain and base of the skull.

"As blood passes through the cooling area, it reduces the temperature throughout the rest of the body," the authors said. "The coolant mist is only used as needed to adjust the body temperature to within the target range."

This cooling process is a means of achieving artificially induced hibernation, or torpor, in the body. Other components of the support system included a venous catheter passing into the chest wall for delivery of the TPN liquid, and a urine collection assembly and drain line. As illustrated in their report, a hypothetical space man, clad only in shorts and fitted with a sleep mask, lay face-up (or stood, there being no difference in zero-g), inside an alcove that somewhat resembled a hospital bed with side rails, straps, and so on, in a well-appointed emergency room complex. Indeed it looked very much like the "hibernation pods" pictured in any number of sci-fi flicks. Inside this alcove, or "Torpor Module," the traveler would spend the approximately 200 days needed to get from Earth to Mars.

In this initial design, there would be six crew members in the habitat, each in his or her own "individual torpor compartment," all of whom would be in hibernation for the full duration of the trip. Their various needs would be taken care of by a robotic manipulator arm.

"The torpor habitat contains two robotic manipulator arms, one overhead and one in the deck, centrally located in the habitat to provide reach to all six crew members," the report said. "The manipulator arms are used to manage and manipulate the crew lines, leads, and restraints during the mission. The arms are redundant; a single arm can access all six crew members."

NASA, however, was not happy with the idea of an entire spacecraft crew being essentially inert and unconscious during the whole of the six-month Earth-to-Mars voyage through space. What if there were complications or emergencies? Shouldn't the astronauts be awakened periodically and be able to stretch their legs and move about?

The SpaceWorks team responded by producing more advanced and bigger designs, including a habitat that would keep a few astronauts awake on a rotating basis. Their "settlement class" design included a 100-person habitat in which 96 passengers would be in torpor while four others would remain awake and active, serving as caretakers to those in hibernation. The habitat would also rotate, providing artificial gravity to the crew, thereby lessening the chance of bone demineralization and muscle loss.

"You get eighty percent of the benefits by cycling through the hibernating crew and waking some up rather than turning out the lights on everybody for six months," John Bradford said of the settlement class scheme.

# # #

Despite the appeal of these design concepts for Mars missions, to date NASA's actual, stated plans for getting humans to Mars, which involve the Space Launch System and the Artemis rocket, do not include putting crew members into hibernation. "There is no NASA baseline design or adoption of the [torpor] capability for currently planned human Mars missions (such as they are)," Bradford said.

Of course the SpaceWorks mission concept has its own limitations, one of which is that the ship still has to carry contingency food stores in case the crew ever ended up being awake and functional. Depending upon how large a mass the emergency food store represents, the ship might end up being much larger than was originally intended.

Separately, the European Space Agency (ESA) was concurrently doing its own research on the use of torpor for deep space missions. In a 2021 paper, "European Space Agency's Hibernation (Torpor) Strategy for Deep Space Missions: Linking Biology to Engineering," a group of six authors affiliated with the ESA laid out the agency's then-current plans for getting astronauts to Mars. While there are many similarities and areas of overlap between the ESA's and the SpaceWorks proposals, one key difference is that the European version did not include a method for continuous feeding of the crew members. Instead, while in the state of torpor, the astronauts would essentially live off of their own body fat, as bears do during hibernation. This body fat is known as subcutaneous white adipose tissue.

"Subcutaneous white adipose tissue (SAT) is considered a very significant source of nutritional storage and for thermal insulation during hibernation," the authors wrote. "In addition, SAT is an endocrine organ that releases hormones including leptin, adiponectin and estradiol." Further, it is known that women tend to make better use of SAT than men, and "this knowledge would suggest women to be preferable candidates for 'manned' deep space missions."

Another difference between the SpaceWorks and ESA mission concepts is that the European scheme allows for periodic awakening of the eight crew members, during which time they would consume normal amounts of food and attend to the other usual activities of the waking state. When either awake or in torpor, the crew members would reside in "hibernacula," spaces that resemble minimalist hotel rooms.

"During non-torpid states, the hibernacula are used as private quarters for sleeping, personal office, private conference and recreation. Those pods provide optimal noise protection and personalized air conditioning and illumination control to ensure ideal torpor conditions," the report said.

For the Europeans, the one big unknown in all of this was how to induce torpor. "Research into how torpor can be invoked in mammals is currently in its initial stages and by no means sufficient for application in humans," the authors stated.

"The induction of torpor in the entire crew imposes a huge demand on the automation of corrective and preventive maintenance. The need to react to mission critical and catastrophic scenarios autonomously or with minimum delay, without ground intervention, imposes very stringent requirements in terms of fault tolerance and probabilistic risk targets."

For the ESA, therefore, a trip to Mars with a largely hibernating crew was far from a sure thing. Which may be just as well, since a year later, in 2022, a group of scientists in Chile published a paper in *Proceedings of the Royal Society B* in which they questioned the notion that long-drawn-out torpor for deep space missions is workable. "The actual amount of [metabolic] savings that hibernation represents, and particularly its dependence on body mass (the 'scaling') has not been calculated properly," the authors said. In fact they suggested that when calculated correctly for heavy animals such as humans, the actual metabolic energy savings from being in a state of hibernation would be so small as to be negligible.

There are a number of takeaways from the studies mentioned above, the first of which is that the maximum duration for which there is actual data about the consequences of being in therapeutic hypothermia, is 14 days. The source for this claim is a Chinese study from the year 2000 in which the patient was described as having suffered severe traumatic brain injury. But the 180 to 200 days it would take to get to Mars is 12 to 14 times longer than 14 days, and it is a major extrapolation to predict anything about the health consequences of putting astronauts in a state of torpor for that far longer amount of time. And that is only for a trip to Mars.

As for an interstellar voyage that could last for 200 *years* or more, any forecast about the likely health effects of being in suspended animation for such a long period is at best an exercise in creative guesstimation. The available data are just not up to the task, and that lack of information puts us in uncharted territory. The effects on the body of complete or periodically interrupted stasis for periods of hundreds of years are unknown but are far more likely to be harmful than they are to be beneficial or neutral. Plus, there is an additional and special reason against putting anyone in suspended animation for hundreds of years, a problem that would not apply to the much shorter duration of a Mars trip. The problem is known as the Hayflick limit.

The Hayflick limit is a finding, in 1961, by Leonard Hayflick, an anatomist then at the Wistar Institute in Philadelphia, and colleague P. S. Moorhead, that normal human somatic cells can divide for only a finite number of times (about 50) before cell division stops. Prior to this, the accepted view, propounded by Alexis Carrell, a French-American surgeon, in 1921, was that normal human cells were immortal, and could divide any arbitrary number of times. But in a series of experiments, Hayflick showed that this was untrue. Cell division stopped eventually, at what is now known as the Hayflick limit, a discovery that had obvious implications for aging, as well as for any hope of human immortality. (Cancer cells, ironically, are not subject to the Hayflick limit, and can continue to divide unstoppably, which is why cancers are so lethal.)

The Hayflick limit also raises a question about using suspended animation for human spacecraft missions that could last for centuries. Do those aboard age normally, or at slower rates? What if they develop cancer? Might the astronauts even die en route? Nobody knows, because there is just no available data one way or the other. Which means that for this reason too, for purposes of interstellar travel, suspended animation is the ultimate untested and unproven technology.

But it is not the only reason. With the crew in stasis for the whole or at least for most of the trip, the task of keeping them alive and well falls to the robots, or to the human caretakers if the trip involves periodic awakening of crew members. But if only a small percentage of the crew is awake and attentive at any given time, it is the robots and sophisticated AI that will see to the needs of the rest of the crew that is asleep. To do this successfully, and to make all the necessary adjustments and repairs, including even possible surgery, these complex robotic systems would have to be paragons of intelligence, diagnostic ability, and manual dexterity.

It is axiomatic that anything that functions may malfunction. And as we have seen, the longer the trip, and the more complex its operation, the more likely it is that malfunctions or failures will occur. The robots themselves will be subject to malfunctions, and some failures may be unrepairable. For this reason also, a failure-prone spacecraft full of hibernating spacefarers does not appear to be an acceptably reliable way of migrating to the stars.

Excepting, that is, in the realm of science fiction; but even there, failures can occur, as indeed happens in the 2016 film *Passengers*. The story is that of a sleeper ship, the *Avalon*, carrying 5,000 colonists and 258 crew members, all of them in hibernation pods keeping them in a state of suspended animation. The ship is on a 120-year-long journey to the extrasolar planet Homestead II. Thirty years into the mission, a stray asteroid strikes and damages the ship, despite its having a sophisticated defense system designed to prevent such collisions.

In consequence of the asteroid strike, one of the hibernation pods malfunctions, opens up, and disgorges a single passenger, Jim Preston, 90 years early. He immediately realizes that unless he can get back into hibernation, he will die aboard the ship, alone. But Jim is a mechanical engineer and tries to restart his hibernation pod so that he can go back into hibernation and then reawaken at the proper time, 90 years in the future. He is kept alive by taking meals at the ship's automat, an automated food-dispensing system.

After trying for more than a year to get back into hibernation, Jim decides he would like a human companion and deliberately wakes another passenger, Aurora Lane, knowing full well that both of them will die on the ship together. He awakens her anyway, and the two end up falling in love. Meanwhile, another pod failure awakens Gus Mancuso, an officer of the ship. He has access to the ship's control system and learns that multiple mechanical failures had occurred following the asteroid strike, including knocking out the ship's fusion reactor power plant.

Mancuso himself soon dies of injuries sustained in his damaged hibernation pod.

But with access to the control room and the power plant, Jim is able to make repairs, after which the ship returns to normal operation. The ship continues on its way, and reaches its destination, but Jim and Aurora die enroute. Ironically, the final success of the trip is owed to the accidental awakening of three crew members who were able to make repairs to a ship that was otherwise headed for disaster. The movie received mixed reviews among critics, one of whom called it a "*Titanic* among the stars."

# # #

Irrespective of whether it's a multigenerational flight or a sleeper ship full of hibernating spacefarers, let us suppose that the craft finally manages to arrive intact. What happens next? If past human experience is to be our guide, then the members of the arriving generation will proceed to re-enact what our human forebears had done across the course of history, which was to, (a) make the world increasingly fit for human habitation, and then, gradually, over time, (b) unintentionally render it increasingly unfit for continued human habitation. This is indeed what we have done to our own native planet, Earth. It is commonplace today, as we observe the phenomena of climate change, global warming, and the emergence of increasingly hazardous weather patterns, that human occupation and use of our resources are putting our very health and lives at risk.

There is no reason to think that matters would be any different on a new planet. "If we cannot solve the problems we have created on Earth," Goldsmith and Rees said in *The End of Astronauts*, "we shall never do so by starting over."

There is ample basis for such a forecast in what the human species has already done to our own celestial close companion, the Moon. By the end of 2012, humans had deposited more than 400,000 pounds of objects, debris, and space trash on the Moon. This included the spacecraft wreckage of the 70 rockets that had landed upon or crashed into the Moon, starting with the 800-pound Russian Luna 2 vehicle in 1959, followed in 1969 by the trail of detritus left behind by the Apollo astronauts, consisting, among other things, of: shovels, rakes, geological tools, lunar experiment packages, reflectors, tongs, scrapers, scoops, lunar landing modules and rovers, boots, television and still cameras, lenses, film magazines, lines and tethers, boxes, canisters, cables, filters, antennas, packing material, hammers, cockpit seat armrests, blankets, towels, wet wipes, personal hygiene kits, empty packages of space food, plus 96 bags of human urine, feces, and vomit. Also left behind were various sculptures, plaques, pins, patches, medals, and miscellaneous other pieces of flotsam, jetsam, and junk (Figure 10.1).

And if history is to be any guide, and supposing any life forms were to be found on the new planet, one of the most pressing tasks of the human space invaders would be to subjugate or

Figure 10.1 The lunar landfill: The Moon's surface as left behind by the crew of Apollo 17, December 1972. (NASA)

otherwise take control of them, or to use them for the colonizers' own purposes. A second immediate task would be to establish a strong defense system that would be sufficient to repel any and all alien intruders. Ultimately, of course, there would be habitats, towns, cities, and all the other manifestations of human occupation and territorial dominion, including a government, courts, police departments, and jails.

As Freeman Dyson said in his 1968 essay, "Human Consequences of the Exploration of Space," "If we succeed in colonizing Mars, Mars will soon resemble the Earth, complete with parking lots, income tax forms, and all the rest."

Hardly the sunny outcome we'd been promised by interstellar travel proponents. And that raises the question: If this is the future of putting the human species on another planet, then why make the extremely arduous, dangerous, and inordinately expensive trip to begin with? In other words, why go?

11 WHY GO?

Aside from the by now all-too-familiar raft of difficulties associated with interstellar travel, there remains the overriding question of why we ought to undertake such a daunting, expensive, and risky enterprise to begin with. Strangely, not everyone agrees that we need a reason. In his 1992 book *From Eros to Gaia*, Freeman Dyson states: "It is natural and right that we shall continue to stumble ahead into space without knowing why."

But this is nonsense. After all, human flight to the stars would be such a huge, hulking, time- and resource-intensive project that it would be warranted only in light of a persuasive and compelling rationale in favor of embarking upon it. As it is, however, instead of such a rationale, the interstellar space literature is overflowing with emotional, romantic, idealistic, extravagant, and metaphorical rhetoric in favor of going to the stars. Such rhetoric, which in the end amounts to no more than a laundry list of hopes, is not to be taken seriously, because hopes, emotional states, metaphors, and dreams have no epistemological status. They are inner states of mind – wishes, feelings, and fantasies – as opposed to accurate representations of reality that are designed to convince the intellect and to instill beliefs that correspond to facts.

But over and above these emotional yearnings there yet remain a series of arguments that space advocates have put forward in the guise of persuasive factual evidence, as hard and strong reasons for believing that human flight to the stars is justified, or is even a necessity. It is ironic, then, that some of those arguments, even when presented as powerful and decisive, turn out to be, upon analysis, by no means sufficient to the task. Indeed, a few of the justifications that interstellar travel enthusiasts have offered are remarkably weak, and they are often opposed by stronger arguments to the contrary. And some of the alleged justifications for interstellar travel are simply beside the point.

As we have seen (in Chapter 6), in 2002 the American Association for the Advancement of Science sponsored, in Boston, a symposium on the topic of multigenerational interstellar migration. One of the speakers on that occasion was Doug Beason, who at the time was deputy associate director of the defense threat reduction division of the Los Alamos National Laboratory. He was the author of a dozen books, including one science fiction novel, and in all possessed an illustrious set of credentials. Beason spoke on the topic "Interstellar Travel: Why We Must Go," which was later published as an essay in the symposium proceedings.

Beason started with an ambitious claim: "This article lays the groundwork for showing the overwhelming need for mankind to embark on interstellar travel." Not just need, but *overwhelming* need. And then asks: "Why bother going to the stars, and why now?" And immediately answers:

> The obvious answer for interstellar travel is that the program would provide a long-term focus for not only our space efforts, but for the nation as a whole. It allows us to invest in our nation's scientific and technological base – traditionally the best product that America has to offer in the post-industrial age era [sic]. This will create new job opportunities and markets in the new economy.

That's it? We should go to the stars for the sake of a "focus," and to "create new job opportunities"? This is not only quite underwhelming; as a case for going specifically *to the stars*, the argument is entirely beside the point for it is directing our attention to benefits that will accrue, here, on Earth, as opposed to the benefits, if any, of traveling to the stars. Moreover, the fact is that investment in any number of other areas – medicine, the environment, affordable housing, better schools – could equally well provide a "focus," and would also create new jobs and markets. In other words, there is nothing specific to interstellar flight that makes it a better than average means of creating jobs and market opportunities. Indeed, if this is the best we can do by way of justification, then clearly we have no reason to go.

But Beason has more to offer. "Further, it would allow for advancing technological innovation." Technology, however, advances on its own, and does so quite apart from any single overarching motivating factor such as starflight. Some of the technological innovations that have occurred since Beason's talk in 2002 are: the introduction of smartphones (as opposed to single-purpose, communication cell phones), self-driving cars, flash drives, completion of the Human Genome Project, electric vehicles, 3D printing, CRISPR gene-editing technology, and the chatbot ChatGPT, among others. And none of these techno-logical innovations occurred as a result of, as a payoff from, or as an outgrowth of an interstellar flight program. Far from show-ing an "overwhelming need" for going to the stars, this argument from Beason does not in fact show *any* need for doing so.

Another speaker at the same symposium was Yoji Kondo, of the NASA Goddard Space Flight Center, who also addressed the question of why we should travel to distant planets. His reason-ing was that "In order to survive as a race, we must get out of the solar system at some point in the future." This claim is distinct from, and far more sweeping than the more usual claim that to survive as a species we must move off of Earth and become a multiplanet species within our own solar system – a point repeatedly made by Elon Musk and reiterated by many others. Still, Kondo was not alone in arguing that we must sooner or later put the entire solar system itself behind us. Other space advocates have made much the same point, based on the theory that some billions of years hence, the Sun will become a red giant, engulfing the planets of the inner solar system, and then become a white dwarf star which, in Kondo's words, "is entirely unsuitable for maintaining human life on Earth or anywhere else in the solar system."

To this there are a number of replies. First, even if it's true that this will happen "over the next few billion years," such an extremely far-off event does not constitute a sufficient reason for our taking any action *now*, literally billions of years ahead of time. Billions of years is an excessively long planning horizon for the intelligent utilization of human time, resources, and effort in the present, when we are already facing a variety of more pressing and immediate existential threats to our health, safety, and security. Additionally, across those billions of years into the future, technology would almost certainly have advanced to a point that is so far beyond what we can accomplish today that it would render current efforts to build a starship primitive and useless as compared to what we could build much later, closer to the time at which an interstellar capability might be needed.

Second, the argument assumes that the human species will still *exist* at the end of the billion-year time frame in question. But this is by no means assured. Among the oldest species known to science are sponges, thought to be 600 million years old, jellyfish, at 500 million years, and cyanobacteria, which are possibly 3.5 billion years old. Modern humans, by contrast, are thought to have developed only some 300,000 years ago, and any number of possible events have the potential to wipe us out long before the billion-year timeframe envisioned by Kondo and others. Climate change, a massive asteroid strike, a global outbreak of a new and deadly viral disease, plus our own susceptibility of being wiped out in a worldwide nuclear weapons cataclysm, all undercut the assumption that the human race will survive intact until that far-off day when the Sun dies.

Species have finite lifetimes; they do not persist forever. Many more species have evolved on Earth and have gone extinct than currently exist, and there is no reason to think that the human species will constitute an exception. Arguably, all of us will someday die individually. And it is entirely possible that, through the action of some currently unknown cause or causes, all of us might perish more or less simultaneously. It is also possible, and indeed more likely, that across billions of years the human race, through natural selection, genetic drift, or deliberate genetic intervention, will evolve or be modified into a species that is no longer human.

# # #

A commonly proposed but equally bad reason for making the trip to the stars (or anywhere else in space) is that we're "wired" to do so, that the need for star travel is somehow embedded in our DNA. Neil Armstrong, the first man on the Moon, expressed this view when he said, "I think we're going to the Moon because it's in the nature of the human being to face challenges. It's by the nature of his deep inner soul. We're required to do these things just as salmon swim upstream." And Robert L. Forward, the science fiction author and futurist, agrees: "Why should we bother going to the stars if it is so difficult? There is one reason that should be obvious to us all. It is built into our genes." Examples of this sentiment could doubtless be multiplied many times over.

The first problem with such a claim, however, is that it is presented without evidence, as if it were obvious and self-evidently true on its face, and therefore it is sufficient merely to make the claim and leave it at that. This amounts to "proof" by mere assertion, essentially by fiat. But to have any significant epistemological status at all, assertions must be backed by facts, by adequate evidence that they correspond to reality. It may be that these and other people who make the same argument – that space exploration "is in our DNA" – are speaking metaphorically rather than literally. In that case the phrase "in our DNA" would be a mere figure of speech. But if the phrase is not meant literally, if the urge to explore and colonize space is *not* literally in our DNA, then of what use is it to the argument? It becomes a mere image or pretty picture, one that is even perhaps poetic in its way. Indeed, the claim that "it's in our DNA" sounds more like an advertising slogan than an actual reason for taking any given course of action. One could even make a little jingle of it, sung to an infectious, repetitive, and annoying tune: "It's in your DNA to drive a Chevrolet." But however well (or poorly) that might work as an advertisement, the one thing it is not is a valid reason for exploring space, much less for going to the stars.

In this vein, author Patrick Lin in a 2006 piece, "Look Before Taking Another Leap for Mankind," asserts that: "Wanderlust, or the compelling need to explore or travel to new places, is in our DNA – that is simply what humans do." But it is not what *all*, or even *most* humans do. Because for every Magellan, Marco Polo, or Captain Cook who has traveled far from home, there are hundreds if not thousands of humans who have never strayed far from their own place of birth. And if we really do have this exploratory instinct, then why are so many of us content just to sit on the couch with our beer and taco chips?

But even if we literally did have a genetic disposition, drive, or urge to colonize space, the existence of such a drive would not by itself *justify* space colonization. From the fact that we have a disposition or urge to do something it does not immediately follow that we ought in fact to satisfy or yield to that urge. Some people may be genetically disposed to violence or aggressive behavior, but having such a disposition does not justify acting upon it and engaging in violence or aggression against others. The fact that we have a given urge merely raises the question of whether we ought to satisfy it; it does not answer the question. Space advocates must show by credible factual evidence that a given space venture has inherent value and yields beneficial consequences, and they must also demonstrate that the value thereby created is not offset or negated by even greater costs.

It is also worth noting in passing that there are genes in the human genome that do not actually get expressed by the organism. For example, inheriting a cancer-related gene doesn't mean that you will necessarily get cancer. It means that your risk of getting cancer is increased. So even if there was a literal gene for space exploration, that gene might not be expressed in all or most members of a given population. And it would not therefore constitute a reason for going to the stars.

# # #

But this is not the end, unfortunately, of bad reasons for going into space. In 1972, three years after the first Moon landing, astronaut Gene Cernan, of Apollo 17, was the 11th, and last man to walk on the Moon. And as he prepared for the final time to climb up the ladder and into the lunar excursion module (the LEM), he said, "I believe history will record that America's challenge of today has forged man's destiny of tomorrow" – his destiny, of course, being to go into space. Separately, Cernan has also said that, "It's our destiny to explore. It's our destiny to be a space-faring nation."

There are endless examples of the sentiment that going to the stars constitutes our "fate," our "destiny." The National Space Society, for example, is on record as saying, "It is human destiny to escape the cradle of our planet of birth." Jane Poynter, the biospherian, explained the motivation for establishing Biosphere 2, by stating: "We were re-creating Eden, saving the Earth and fulfilling mankind's destiny to reach the stars." And so on.

Apollo astronaut Buzz Aldrin, the second man on the Moon, was positively effusive on the subject, writing (with coauthor Wyn Wachhorst) in "The Urge to Explore" that:

> Like the sailing ships that incarnated the aura of the Renaissance, or the great steam locomotives that embodied the building of America, the Apollo rocket is an emblem of the *human* spirit. Apollo was inevitable from the first gleam in the eye of the hunter-gatherer, from the first fire, wheel, and furrow; it was latent in the stirrup and the longship, in the creak of every caravel, the ring of every railroad spike, the lonesome howl of every lumber camp harmonica. From the moment the first flint was flaked, space was *fated* to be the final canvas for expressing in bold strokes the inexhaustible soul of humanity.

Wow. Such florid rhetoric notwithstanding, neither fate nor destiny has any real existence, and therefore can play no valid part in guiding human choices or actions. Philosopher James S. J. Schwartz argues further that concepts such as destiny and fate are mystical or supernatural entities, and as such are "unscientific and should play no role in scientifically-minded space advocacy."

In a different vein, historian Patricia Limerick observes, "Commit yourself to a *destiny*, and you are handing over your free will; you are *volunteering* for compulsion; you are doing things because you *must* do them, not because you have reflected, pondered, and chosen to do them."

Arguments to the effect that we go into space because it's in our DNA or is our destiny stand in stark contrast to the *actual* reason that we in fact did go into space during the Apollo era. As is widely recognized among historians of the space program, the actual reason the United States went to the Moon in 1969 was not the same as the stated reasons. The *stated* reasons were expressed by President John F. Kennedy in a speech that he delivered at Rice University in Texas, on September 12, 1962. In an oft-quoted passage, he said:

> We choose to go to the moon. We choose to go to the moon in this decade and do the other things, not because they are easy, but because they are hard, because that goal will serve to organize and measure the best of our energies and skills, because that challenge is one that we are willing to accept, one we are unwilling to postpone, and one which we intend to win, and the others, too.
>
> It is for these reasons that I regard the decision last year to shift our efforts in space from low to high gear as among the most important decisions that will be made during my incumbency in the office of the presidency.

As memorable and stirring as those words may have been, they did not in fact embody Kennedy's real motivation in putting the American space program into "high gear." He did so in the context of the Space Race, which was a battle between the Soviet Union and the United States for supremacy in space. The Soviets were the first to put a satellite in orbit, on October 4, 1957. The very next month, on November 3, 1957, they put a dog, Laika, into space. Clearly, the Soviets were getting some-where, and leaving the Americans in the dust. And then less than four years later, on April 12, 1961, the Soviets put cosmo-naut Yuri Gagarin into Earth orbit, on a flight that lasted 108 minutes. That was the last straw for John F. Kennedy.

Kennedy could not let the Soviets continue to outpace us, and therefore proposed a grand new goal for the United States: to be the world's first nation to put a man on the Moon. So in short, the actual reason the USA went to the Moon in 1969 was simply to beat and best the Russians. This was why time was so much of the essence, as it had been even before Kennedy's Rice University speech, for as early as October 1958, NASA had set forth the objectives of its manned satellite program in a 2½-page typescript that stated:

I. Objectives

The objectives of the project are to achieve at the earliest practicable date orbital flight and successful recovery of a manned satellite, and to investigate the capabilities of man in this environment.

At the earliest practicable date. This marked the start of Project Mercury, whose initial, suborbital flight was made by astronaut Alan Shepard on May 5, 1961. Fate, destiny, and DNA had nothing to do with it. The goal was specifically to beat the Soviets to the Moon, and that objective was achieved on July 20, 1969. Today, however, there exists no comparable race to the stars, and so Kennedy's motivation for going to the Moon is no longer alive and operative. Which again leaves us with the question: Why go?

# # #

Another bad answer is: Because it's our moral obligation to do so.

That answer, admittedly, is not often advanced in the litera-ture of interstellar exploration, but it does have its defenders. Christopher E. Mason, whose book *The Next 500 Years* we have already encountered (in Chapter 5), claims that human beings have a "duty to engineer," and that, more particularly, we have a moral obligation to ensure "the survival of as many of the life-forms [on Earth] and molecules as possible."

In laying out his case, Mason proposes "a new kind of ethics," which he calls "deontogenic ethics":

Deontogenic ethics is based on two simple assumptions. First, assume that only some species or entities have an awareness of extinction. Second, assume that existence is essential for any other goal/idea to be accomplished – in short, *existence precedes essence*. Therefore, to accomplish any goal or idea, sentient species (currently humans) need to ensure their own existence and that of all other species that enable their survival. Any act that consciously preserves the existence of life's molecules (currently nucleic acid-based) across time is ethical. Anything that does not is unethical.

But this is poor reasoning. Mason's argument is that since human beings are aware of extinction, and since entities must exist in order for goals to be accomplished, it follows that humans must ensure their own existence and that of all other species that keep humans alive. It might be thought that there must be more to Mason's argument than this, but there isn't. Indeed, when he restates his argument a few pages later, it amounts to the same thing: "We know that humans are still the only species with the knowledge of extinction, and thus, we have a duty of stewardship of our own species and that of others."

But from the fact that humans are the sole species to be aware of extinction, it does not follow that humans have any sort of duty to keep themselves or other species alive. What's wrong with such a claim is something very basic, namely that in order for an argument to be logically valid, nothing can appear in the conclusion that is not contained in the premises. The standard Logic 101 example of a logically valid inference is the syllogism (which goes back to Aristotle):

Socrates is a man.
All men are mortal.
Therefore Socrates is mortal.

Mason's argument does not even pretend to be a syllogism. Formally expressed it amounts to this:

Humans alone are aware of extinction.
Entities must exist in order for any goals
to be accomplished.

Therefore humans have a duty to ensure the survival of themselves and other species by going into space.

That is a complete non sequitur: what's wrong with it is that the concepts of "duty" or what's "ethical" do not appear in the premises. They just materialize out of nowhere. It may be true that Mason sincerely believes that humans have such a duty, and he may even believe that the inference expressed above is valid, but that doesn't make it so. These beliefs are just his *opinions*, but we don't go to the stars on the basis of mere opinions.

A more plausible, but still not decisive argument that we have a moral obligation to go to the stars is given by the philosopher James Schwartz. In "Our Moral Obligation to Support Space Exploration" (2011), Schwartz states that:

The obligation to support space exploration can be defended in at least three ways: (1) the "argument from resources," that space exploration is useful for amplifying our available resources; (2) the "argument from asteroids," that space exploration is necessary for protecting the environment and its inhabitants from extraterrestrial threats such as meteorite impacts; and (3) the "argument from solar burnout," that we are obligated to pursue interstellar colonization in order to ensure long-term human survival.

However, even if we accept all three propositions, that space exploration will give us access to asteroidal and other resources; will allow us to defend ourselves against meteorites (by intercepting or destroying them); and finally that interstellar colonization might be useful in saving us from solar burnout, it does not follow that we have an *obligation* to do any of those things. What follows is that we have reasons to take those actions as practical measures that will bring about the ends in question. But no *obligation* per se arises from the fact that those measures will be helpful in attaining those ends.

The relationship between means and ends is merely a practical matter, not a matter of moral obligation. In general, obligations arise when a rational agent promises, agrees, or vows to produce a given end-state. For example, when parties to a marriage each vow to remain monogamous throughout the marriage, then an obligation exists for both parties to remain faithful. When parties to a treaty pledge to take a certain course of action, they are then under an obligation to take that action. And when a loan applicant promises to repay a loan, then that applicant is under an obligation to do so. But absent such promissory declarations, no obligation exists. And that is the case here: going into space may be a means of accomplishing a certain goal, but the fact that it is a means does not imply or entail that there is a moral obligation to go into space or to actually accomplish that goal.

Another way or saying this is that going into space has instrumental value in bringing forth a particular result. But the fact that it has such an instrumental value does not imply that anyone is under a moral obligation to produce that value or to bring about that result.

So in this case, too, therefore, the desired conclusion does not in fact follow from the initial premises, and so Schwartz's three arguments fail. But for better or worse, there are yet more bad arguments to come.

# # #

Off in its own curious corner of wild improbability is an offbeat idea of Carl Sagan's. In 1963, Carl Sagan, who was at that time still in his twenties and a newly appointed member of Harvard University's astronomy department, presented a novel reason for star travel: to discover and make contact with alien intelligent civilizations. In "Direct Contact among Galactic Civilizations by Relativistic Interstellar Spaceflight," Sagan cast doubt on the notion that communication with aliens by electromagnetic means was a realistic possibility. "The difficulties of electromagnetic communication over such [large] interstellar distances are serious," he said. "A simple query and response to the nearest technical civilization requires periods approaching 1000 years." An extended conversation would be nearly impossible. Second, he noted that there is no agreement even among Earth's space scientists as to what radio signal frequency we ought to monitor when conducting the search. Finally, electromagnetic communication would not allow for "two of the most exciting categories of interstellar contact – namely, contact between an advanced civilization and an intelligent but pre-technical society, and the exchange of artifacts and biological specimens among the various communities."

Sagan's solution was straightforward: just go out there and meet them. The vehicle he proposed for making the trip was the then new, but now long since discredited, Bussard Interstellar Ramjet. But surely this is putting the cart before the horse. How do we know which destinations to visit unless we first know that there's an advanced civilization out there awaiting us? Sagan sidesteps this issue by assuming the existence of so many starships moving at such great relativistic velocities that they will just stumble across other civilizations "on a purely random basis." "The number of contacts per year for the Galaxy as a whole is then 10^6; a sizeable fraction of these should be between two advanced communities."

Sagan's proposal makes so many unlikely assumptions as to make it difficult to take his argument seriously. For one thing, he assumes an arbitrarily large number of starships: "The mean number of starships on patrol from each technical civilization at any given time is $\sim 10^3 - 10^4$." But this is pure fiction, an ad hoc number pulled out of a box. He assumes that an advanced civilization will be recognized as such when it is encountered. And he assumes that some sort of communication between the two civilizations will be possible. Otherwise, what is the meaning of "contact" if the two civilizations cannot communicate but can only stare at each other across the void?

But the whole idea of communicating with an alien species is itself problematic since there is no common language shared by humans and putative extraterrestrials, nor is there any proof that it would be possible to construct such a lingua franca in a way that would make it intelligible to members of both cultures. Basically, this is the problem of communicating meaning or knowledge outside of our current cultural and intellectual context. The difficulty involved can be gauged by the fact that there exist artifacts on Earth created by humans from the ancient, or even relatively recent past, that remain undeciphered even after hundreds of years of study: the Voynich manuscript, for example, an undecrypted, illustrated codex from the early fifteenth century. Scholars disagree as to what it says or what it means, despite its having been created by one or more members of our own human race. These difficulties are only multiplied exponentially when The Others are *not* human.

In the late 1950s, Dutch mathematician Hans Freudenthal created a language intended for human use in communicating with aliens: *LINCOS: Design of a Language for Cosmic Intercourse*. The language was supposed to be self-interpretable, but the meaning of even the simplest messages is not clear without an appended explanation, which, of course, is no more intelligible than the message it is intended to explain. An example conversation is given in Table 11.1.

Table 11.1 Example of a conversation in the LINCOS language.

LINCOS text	Meaning
Ha Inq *Hb* ?x 4x=10	Ha says to Hb: What is the x such that $4x = 10$?
Hb Inq *Hc* ?y y Inq *Hb* ? x 4x=10	Hb says to Hc: Who asked me for the x such that $4x = 10$?
Hc Inq *Hb* *Ha*	Hc says to Hb: Ha.

Ha, indeed!

In light of these multiple and substantial difficulties with Sagan's proposal, it does not constitute an acceptable rationale for traveling to the stars.

# # #

In contrast to Sagan's flawed reasoning, perhaps the most widely accepted and superficially plausible reason for interstellar voyaging invokes the concept of "crossing the final frontier." In their recent (2023) book, *A City on Mars*, authors Kelly and Zach Weinersmith present the so-called frontier thesis in this way:

> The claim is that the United States became dynamic, democratic, ruggedly individualistic, and generally awesome due to a longstanding frontier culture. Sometimes this is a simple rhetorical flourish about space as a place of newness and adventure, but often the frontier is seen as something more – as a process of social resurrection. In this vision, space settlers will forge a hard, serious, creative civilization, and that borderland society will show Earth people a tougher and more democratic mode of life, just as the American West purportedly did for the United States.

The frontier thesis is associated with historian Frederick Jackson Turner and his essay, "The Significance of the Frontier in American History," which goes all the way back to 1893, when he delivered it as a talk before a meeting of the American Historical Association, in Chicago. Turner's central claim is contained in a key sentence: "The existence of an area of free land, its continuous recession, and the advance of American settlement westward, explain American development." By parity of reasoning, then, space advocates emphasize the importance of space as the new, and perhaps final frontier for the further advancement of the human species. After all, space is open in every direction (although transiting it is expensive), there is always more of it, and so we ought to be making this epic journey across it to the stars.

One problem with the frontier thesis is that it has been generally rejected by contemporary, mainstream historians. In a 1987 essay on "Revisiting the Vanishing Frontier," historian William Cronon wrote that, "In the half century since Turner's death, his reputation has been subjected to a devastating series of attacks which have left little of his argument intact."

But even if the frontier thesis has been rejected by historians for technical historiographical reasons, the essence of the concept remains unchanged on the purely intuitive, commonsense basis that we always want to know what's on the other side of the hill, or what's just over the horizon. The unknown still exerts its pull. But why should we yield to that pull? The frontier thesis is in essence the "it's in our genes" concept in disguise: it embodies the idea that exploration is a natural, inborn tendency, that we have a natural urge to explore. And as such it is vulnerable to the same objection: Why should we satisfy that urge? Is crossing the final frontier really going to have only benefits? No costs, no downsides?

So even if we accept the concept of space as the final frontier as a valid and meaningful category, there still exists the problem of a reason for crossing that frontier and for traveling to the stars. And thus far we have seen no convincing, logically valid argument that we should do so.

# # #

The distinct lack of so much as a single persuasive argument for human interstellar travel suggests that the enthusiasm bordering on giddiness for the idea exhibited by many of its proponents might be more a matter of messianic, religious faith and evangelical space fervor than a result of rational analysis and cautious deliberation. Some of the zealousness demonstrated by space proponents also has an esthetic dimension, as if going into space was akin to experiencing the sublime. For example, there is the iconography of astronauts ascending heavenward from the Earth on great, bright plumes of flame. Astronaut Michael Collins's book about his participation in the Apollo 11 Moon mission was called *Carrying the Fire*. In his book, Collins writes of the Apollo program that, "Of course, Apollo was the god who carried the fiery sun across the sky in a chariot." There is implicit in the experience of rising heavenward the notion that a great, almost godlike power is lofting these astronauts atop limitless, awesome, cosmic bursts of energy.

Charles Lindbergh wrote a foreword to *Carrying the Fire* in which he described the effect of seeing the ascent of the Apollo rocket as an almost mystical, transcendent experience:

> You must have witnessed an Apollo ascent to know the title's aptness. When a countdown at Cape Canaveral touches zero, you think the giant rocket will be consumed by bursting clouds of flame and thunder. When I stood with astronauts three miles away from Apollo 11's pad, my chest was beaten and the ground shook as though bombs were falling nearby. Then a flame arose, left the ground behind – higher, faster, a meteor streaking through the sky. It seemed impossible for life to exist while carrying that ball of fire. Yet it did exist, as I saw later on a television screen that portrayed weightless men, countless instruments, and planet earth through a window.

The full launch process constitutes an epiphany for all concerned: the astronauts, the launch controllers, the on-site spectators, and those at home or at work, watching all of it unfold on TV. Some of those spectators feel a wave of awe, others are choked up and moved to tears, all are deeply affected by the incredible light and sound show in front of them.

Space zealots, for their part, regard astronauts as heroes, practically saints, despite all the hijinks of the Mercury Program astronauts as depicted by Tom Wolfe in *The Right Stuff*, which talked about their "flying and drinking and drinking and driving," the serial adultery, their lying to the medical staff, all of that. But then, afterward, when the Moon had been conquered, the astronauts were practically worshipped as gods as they rode through miles of ticker tape parades, gave addresses before

Congress, were splashed across the pages of *Life* magazine, did television interviews, gave press conferences, were summoned to the White House, and more.

There is in fact a small scholarly literature in defense of the idea that going into space is more a matter of religious faith than a legitimate human need. Roger D. Launius, former chief historian for NASA, and a curator at the National Air and Space Museum, argues in "Escaping Earth: Human Spaceflight as Religion" (2013), that:

> All the elements of religion are present among those who advocate for aggressive space exploration. The belief system has saints, martyrs, and demons; sacred spaces of pilgrimage and reverence; theology and creed; worship and rituals; sacred texts; and a message of salvation for humanity as it ensures its future through expansion of civilization to other celestial bodies. ... It also involves both prophetic and priestly functions; through the prophetic voice it calls on adherents to go beyond the mundane of workaday life to embrace and magnify a higher calling, while priestly aspects serve to establish norms and orthodoxy in belief and action.

The space religion promises salvation and redemption in the form of a new beginning for humankind, to be located somewhere else in the cosmos, off among the stars and the planets, where we will achieve a utopian near-perfection on a pristine world, combined with the prospect of immortality for the species. The problem, however, according to Launius, is that "statements of humanity's salvation through spaceflight are fundamentally statements of faith predicated on no knowledge whatsoever. We have no idea whether or not humans may avoid extinction in this manner. ... Absent the discovery of an Earthlike habitable exoplanet to which humanity might migrate, this salvation ideology seems problematic, a statement of faith rather than knowledge or reason."

The spaceflight religion embodies a new clerical caste (engineers and astronauts), arcane rituals (mission control countdowns and holds, checklist prompts and readbacks), and a language of devotion (NASA-speak jargon). The religion has its revered leaders (astronauts and cosmonauts), martyrs (those who have died aboard spacecraft), and the ultimate prelaunch ritual.

> The process whereby the astronauts prepare for their missions is highly ritualized. It consists of arrival at the Kennedy Space Center in Florida and isolation for some time in preparation, a ritualistic breakfast, suiting up in a special facility complete with oversized stuffed chairs for the astronauts to lounge, the "walk out" to the transport vehicle to travel to the launch site, riding the elevator to the top of the gantry, and then entering the spacecraft from the "White Room," where they are strapped in.

Further on this same theme, the religion of spaceflight has its sacred places, first in the form of launch sites to which people make pilgrimages, often traveling hundreds of miles in order to do so, enduring long waits and sometimes repeated postponements, or outright cancellations of the flight. Secondary sacred spaces include shrines such as the National Air and Space Museum in Washington, DC, which houses iconic objects such as space capsules, space suits worn by astronauts (the ecclesiastical vestments), and other such sacred gear and paraphernalia.

Finally, adherents of the spaceflight religion are united by a strong group identity centered around the core belief that humankind must go into space in order to achieve a more perfect spiritual and physical salvation among the stars, as well as to attain the state of species immortality. (If there is a bible of the space religion, a sacred text, it is *JBIS*, the *Journal of the British Interplanetary Society*, avidly read by the interstellar faithful.) The only important element lacking in the space gospel is an intelligible rationale supporting the core creed.

It is tempting to think that this entire line of argument is itself not to be taken seriously, that it is at best a spoof or satire. And while it is easy to find some humor in the concept of a spaceflight religion, the idea yet embodies a grain of truth inasmuch as it makes it all too clear that the core dogma, that humans must leave the Earth and colonize the stars, exists without a plausible foundation. Without exception, all the arguments we have considered in favor of that viewpoint lack an adequate underpinning in reason and logic. It may be that some other argument, or group of them, provides a sufficient and solid basis for the dogma. But equally, there may be no such justification or proof at all, now or in the offing. And in that case, the whole idea would remain just as much of a dream as it ever had been.

12 THE ODDS

Interstellar travel may be a dream that we ought to relinquish, no matter how disappointing and disillusioning it might be to do so. For while dreams motivate and inspire us, they can equally well lead us astray and entice us to take actions that are unwise and reckless.

In the settings of interstellar travel and human migration to the stars, the dream alone is not a sufficient basis for action. "Dream-walking into space is sleepwalking into space," says Daniel Deudney in *Dark Skies*. "Of the many domains of choice demanding prudent decisions and sober restraints, space thinking is particularly prone to dreamy assumptions, beguiling illusions, and stark disorientations."

Substituting emotional states, longings, and yearnings for the blunt facts of reality in the context of an enterprise as large, complex, and imposing as interstellar voyaging is hazardous to human life, health, and happiness. The reason is that for all of its emotional appeal, magnetic charm, and aesthetic attractiveness, the dream of star travel counts for nothing cognitively. That is to say, the dream by itself tells you nothing about the authentic nature of the undertaking, about what is actually possible, practical, or desirable. It tells you only about what you may feel or want, as opposed to what exists or what plausibly might exist.

Many of the proposed design concepts encountered in our survey of various interstellar vehicles are remarkably far-out: a world ship carrying 700,000 people; another that's 2,865 miles long and capable of carrying 99 billion people; and a laser-pushed starship that requires for its operation an amount of energy that is equivalent to 75,000 times the total power output of the entire world. Such over-the-top, unhinged designs are so much the products of juvenile aspiration, magical thinking, systematic denial, and sheer silliness that it is surprising to see them appear in print, in highly regarded technical journals.

The environment of interstellar space is desolate, harsh, and inhospitable – a vast version of nowhere. It is radically unlike, and far more menacing, than any location that exists on Earth. It is characterized by extremes of temperature and by an absence of a breathable atmosphere – indeed of *any* atmosphere. It is an almost complete vacuum, but is nevertheless continuously penetrated by intense radiation and by rapidly moving particles that are harmful to biological structures.

There are no rest stops, truck stops, sanctuaries, or oases between Earth and the planets of any distant solar system. There is no mid-journey location where interstellar travelers could obtain any needed repairs, supplies, or assistance. For these reasons, one does not enter this sterile, hostile, intimidating realm on the basis of hopes, imaginings, or illusions. What we need, instead, is a severe and sober calculation of the odds.

Historically, as we have seen, all too many dreams have gone up in smoke, starting with the Icarus of legend who plunged into the sea and drowned. Then, during the Middle Ages, various men with dreams of flight put together wings made from wood, feathers, and cloth, and jumped from rooftops, bridges, and towers. Wings flapping ineffectively, they tumbled to their deaths. Eilmer, a monk at Malmesbury Abbey, was the rare exception.

The Wright brothers of course succeeded in getting airborne, but they proceeded step by step and scientifically, experimenting first with kites, and then with unmanned gliders, and then with gliders that they themselves flew. The Wrights also had a proof of concept in birds. There is no proof of concept for a starship.

In sum, the prospect of human travel to the stars faces such an exceptionally wide and diverse assortment of obstacles, improbabilities, multiple risks, and inestimable costs, as to make any attempt to traverse the final frontier far more likely to end in tragedy than to succeed in getting human beings safely lodged on the surface of an extrasolar planet that is in all respects suitable for continued and sustained human life.

There are, in general, seven separate categories of problems facing starflight: physical, biological, psychological, social, financial, ethical, and motivational. Starting with the physics of the enterprise, we have seen that none of the three icons of star travel embodies a realistic, practical, proven design that would be likely to work as advertised. Not the nuclear-powered Bernal sphere, nor the Bussard Interstellar Ramjet, nor the Project Daedalus rocket, which in any case was not even intended to carry passengers. Project Orion represented the high-watermark of deep space craziness, as many project members themselves realized afterward. As Freeman Dyson acknowledged much later, "We really were a bit insane, thinking that all these things would work." Amen.

The world ship, or multigenerational starship, is a uniquely attractive concept. Supposedly, it would carry us and all we needed by way of supplies to take us to wherever in the Galaxy we wanted to go. But there are if anything too many designs for the ship's architecture rather than too few, and no agreement among their proponents as to which if any of them is likely to be workable, nor upon the size of the crew, or the type of propulsion system that could propel such a craft through space. Moreover, a truly earthlike exoplanet has never yet been found, whereas any that might be discovered could be excessively distant, making a voyage to it unrealistically prolonged.

Many of the propulsion system designs that exist are very much of the Hail Mary variety – highly unlikely to work, although not impossible. Controlled nuclear fusion is under active development, and the National Ignition Facility had a few successes that yielded momentary fusion reactions, but never achieved anything even remotely close to a net power gain. Antimatter reactions have also been produced in the laboratory, but antimatter exists in such minute quantities as to make the reaction unusable for interstellar drive purposes. Laser-beamed propulsion, for its part, has multiple problems of its own.

Perhaps worst of all is the very serious problem of mechanical breakdowns aboard the ship, especially a malfunction of the craft's automated self-repair system. Mechanical structures are inherently failure-prone and can be counted upon to fail across an extended time in use. Nothing works perfectly 100 percent of the time, and it is not in the nature of real-world events that any given component of an interstellar spacecraft, much less all of them, would be exempt from the inexorable influence of wear and tear, constant use, and the equally inexorable law of entropy.

Biologically, there are serious health consequences of going into space, particularly in the form of radiation damage to human genes. Such damage can be substantially attenuated by shielding, but not entirely eliminated, due to the generation of secondary particle emissions. Christopher Mason has put forward an ambitious plan to refashion the human body so that we could live even on planets that were not as earthlike as Earth itself. This is an imaginative proposal, but neither a credible nor an appealing one, and in any case would most likely be unpopular to the general public, who would see it as yet another instance of scientists "playing God." His overall genetic redesign proposal would result not in a better human, but in a race of green, bug-eyed, photosynthesizing, post-human monstrosities. And in any case, a program to radically alter the human makeup is bound to be deemed illegal and politically untenable.

Suspended animation, for its part, does not appear to be feasible, especially given the long time horizon of an interstellar trip. The prospect of human reproduction in space is as yet an unknown variable. And above and beyond these issues, one of the greatest health hazards of them all is the possibility of high-speed collisions with interstellar matter, conceivably resulting in the destruction of the ship and the extinction of its crew.

On the positive biological side, the Biosphere 2 ecosystem experiment was generally successful, except for continuous oxygen loss. But it otherwise demonstrated the viability of maintaining human life for sustained periods in an artificial habitat.

Psychologically and socially, there are the ever-present dangers of crew boredom during prolonged stays in space, plus conflict and factionalism developing among crew members, along with sleep disturbances, cognitive impairments, and depression.

Ethically, despite initial appearances to the contrary, there doesn't seem to be any moral problem in launching a multigenerational trip so long as a humane, stimulating, and diverse environment is provided to the travelers.

Financially, the cost of an interstellar voyage is incalculable at present in light of the fact that none of the key parameters such as ship size and type, its propulsion system, destination, trip duration – none of these are known even to a first approximation. Nevertheless, even if and when these factors did become known, a cost estimate, no matter how huge, is likely to be an underestimate due to the inherent complexity and uniqueness of the spacecraft and the multiplicity of its systems and their interrelated components. Very probably, the cost of making such a trip is likely to be extremely high, possibly prohibitively so, whereas the needed resources could be better spent on improving human life here on Earth.

And, as for the final category of motivation for making the trip to begin with, we have found, to some surprise, that of the many different reasons that proponents have offered for making a journey to the stars, not one of them withstood scrutiny. In short, we have found no good reason to go.

So, what are the odds, in the light of all these negatives, even when ameliorated by two positives, that an interstellar voyage would be successful? The only possible answer is that the odds are so heavily stacked against it that we should abandon this dangerously seductive otherworldly delusion.

In 2015, the year of publication of Kim Stanley Robinson's novel *Aurora*, the author wrote an essay, "Our Generation Ships Will Sink," that undertook a systematic assessment of the prospects of a multigenerational voyage to the stars being successful. After an analysis much like that given above, his conclusion was, "Multigenerational starship travel is simply very, very, very unlikely to succeed. If the odds are something like a million to one, should we try it? Maybe not."

# # #

Finally, in 1967, 10 years after *Sputnik I* and two years before the Apollo 11 Moon landing, science fiction writer Arthur C. Clarke put together a collection of readings about rocketry and space travel, calling it, *The Coming of the Space Age*. One of the collected papers was "Cristobal Columbo," by Ralph S. Cooper, a Los Alamos scientist. The piece was a parody, and constituted a fake but generally skeptical report of the Senate of Genoa, apparently delivered in the year 1492, concerning a proposal by one Christopher Columbus to pioneer a new sea route to the Orient.

The piece embodies what may be called the "Christopher Columbus objection" that is sometimes leveled against starship critics: that their skeptical assessments would have prevented the voyages of Columbus. But there is really no analogy here, for a voyage across an ocean on Earth is hardly comparable to a voyage across interstellar space to a solar system light years away. For one thing, Columbus had no need to design and build a radically new type of seagoing vessel: he used a tried and true type of sailboat that already existed, a three-masted caravel, a type that had been in continuous use for decades past. For another, he also did not have to invent a new propulsion system: he used the wind, for free. Third, he did not have to bring, contain, maintain, and replenish his own atmosphere: it existed inescapably all around him, also for free. And fourth, Columbus did not in fact achieve his goal, which was to reach Cathay (China). According to historian David Abulafia in *The Boundless Sea* (2019), Columbus never realized where it was that he actually ended up: "At no stage did Columbus express serious doubts that he had reached Asia." But he hadn't.

There is a consolation to relinquishing a dream which, when viewed realistically and without rose-colored glasses, turns out to be more like a nightmare than a pleasant or entertaining version of the future. And this is that it allows us to realize how fortunate we are to be living on an already suitable (because we evolved here) Planet A, which is a very earthlike Earth at that, arguably far more so than any other likely to be found anywhere else in the cosmos. And, as a corollary, that our one best hope for the survival of our species is to protect, preserve, and maintain it.

BIBLIOGRAPHY

Preface

Clarke, Arthur C. (1964). *Profiles of the Future*. New York: Bantam.

Gingell, Tom W. (2012). "Starship Collision Warning Using Quantum Radar." *100YSS 2012 Public Symposium*, Houston, TX, September 14, 2012.

Goldsmith, Donald, and Rees, Martin. (2022). *The End of Astronauts*. Cambridge, MA and London: The Belknap Press of Harvard University Press.

Millis, Marc. (2010). "First Interstellar Missions, Considering Energy and Incessant Obsolescence." *Journal of the British Interplanetary Society*. Vol. 63: 434–443.

Obousy, Richard. (2012). "Starship by Design: An Exploration of Cutting Edge Interstellar Technologies." *100YSS 2012 Public Symposium*, Houston, TX, September 14, 2012.

Odenwald, Sten. (2022). Email to Ed Regis, April 13.

Chapter 1: Origins of the Dream

Anderson, John D., Jr. (1997). *A History of Aerodynamics and Its Impact on Flying Machines*. Cambridge, UK: Cambridge University Press.

Burroughs, Edgar Rice. (1963 [1912]). *A Princess of Mars*. New York. Ballantine Books.

Deudney, Daniel. (2020). *Dark Skies: Space Expansionism, Planetary Geopolitics, and the Ends of Humanity*. New York: Oxford University Press.

Hallion, Richard P. (2003). *Taking Flight: Inventing the Aerial Age from Antiquity Through the First World War*. New York. Oxford University Press.

Kekulé, Friedrich August. (1958). "The Valency of Carbon and the Structure of Benzene." In Hurd, D. L. and Kipling, J. J. (eds.). *The Origins and Growth of Physical Science*, Vol. 2. Baltimore: Penguin Books.

Kepler, Johannes. (1634). *Somnium*. https://somniumproject.wordpress.com

Lowell, Percival. (1908). *Mars as the Abode of Life*. New York: Macmillan.

Lucian. (2nd century AD). *Trips to the Moon*. The Project Gutenberg eBook. Edited by Henry Morley, Translated by Thomas Francklin. https://www.gutenberg.org/ebooks/author/1997

Sagan, Carl. (1980). *Cosmos*. New York: Random House.

Tsiolkovsky, K. E. (1898). "Reactive Flying Machines." In *Collected Works of KE. Tsiokovskiy*. Volume ZZ. Blagonravov, A. A, Editor in Chief. Translation of "K.E., Sobraniye Sochmeniy, Tom I- Reaktivnyye Letatal'nyye Apparaty."

Moscow: Izdatel'stvo Akademii Nauk SSSR. 1954. NASA 1 T F-237, 1965, pp. 72–117. https://www.tsiolkovsky.org/wp-content/uploads/2021/10/288-tsiolkovsky-collected-works-volume-2-reactive-flying-machines-english-1948.pdf

Chapter 2: The 100 Year Starship

Alcubierre, Miguel. (1994). "The Warp Drive: Hyperfast Travel Within General Relativity." *Classical and Quantum Gravity*. L73–L77.

Angelica, Amara D. (2010). "NASA Ames' Worden Reveals DARPA-funded 'Hundred Year Starship' Program." https://www.thekurzweillibrary.com/nasa-ames-worden-reveals-darpa-funded-hundred-year-starship-program

Cleaver, Gerald B. (2012). "Spacecraft Propulsion via Chiral Fermion Pair Production from Parallel Electric and Magnetic Fields." *100YSS 2012 Public Symposium*, Houston, TX, September 14, 2012.

DARPA. (2010). "DARPA/NASA Seek to Inspire Multigenerational Research and Development." News Release. Arlington, VA: DARPA.

DARPA. (2011). "Request for Information (RFI) 100 Year Starship Study." https://spaceref.com/status-report/darpa-request-for-information-100-year-starship-study/

Dickson, Paul. (2007). *Sputnik: The Shock of the Century*. New York: Walker Books.

Kakaes, Konstantin. (2013). "Warp Factor." *Popular Science*. April. https://www.popsci.com/technology/article/2013-03/warp-factor/

Millis, Marc. (2011). "100 Year Starship Meeting: A Report, by Paul Gilster." https://www.centauri-dreams.org/2011/01/28/100-year-starship"meeting-a-report/

National Research Council. (2014). *Pathways to Exploration: Rationales and Approaches for a U.S. Program of Human Space Exploration*. Washington, DC: National Academies Press.

Obousy, Richard. (2012). "Starship by Design: An Exploration of Cutting Edge Interstellar Technologies." *100YSS 2012 Public Symposium*, Houston, TX, September 14, 2012.

100YSS. (2012). "2012 Public Symposium. 'Transition to Transformation: The Journey Begins." https://www.100yss.org/symposium/2012

100YSS. (2017). "LOOK UP." Press Release. September 15. https://www.100yss.org/news/press

Overbye, Dennis. (2011). "Offering Funds, U.S. Agency Dreams of Sending Humans to the Stars." *New York Times*, August 17.

White, Harold (2011). "Warp Field Mechanics 101." *100YSS Public Symposium*. Orlando, September 2011.

White, Harold. (2012). "Warp Field Mechanics 102." *100YSS 2012 Public Symposium*, Houston, TX, September 14, 2012.

Chapter 3: Three Icons of Star Travel

Bernal, J. D. (1929). *The World, the Flesh, and the Devil: An Enquiry into the Future of the Three Enemies of the Rational Soul.* Foyle Publishing. https://www.quarkweb .com/foyle/WorldFleshDevil.pdf

Bond, Alan. (1974). "An Analysis of the Potential Performance of the Ram Augmented Interstellar Rocket." *Journal of the British Interplanetary Society.* Vol. 27: 674.

Bond, Alan, and Project Daedalus Study Group. (1978–1981). *Project Daedalus: The Final Report on the BIS Starship Study,* 2nd ed. London: British Interplanetary Society.

Bussard, Robert W. (1960). "Galactic Matter and Interstellar Flight." *Astronautica Acta.* Vol. 6: 179–194.

Cleator, Philip E. (1936). *Rockets Through Space: The Dawn of Interplanetary Travel.* London: George Allen & Unwin.

Deudney, Daniel. (2020). *Dark Skies: Space Expansionism, Planetary Geopolitics, and the Ends of Humanity.* New York: Oxford University Press.

Fishback, John Ford. (1969). *"Relativistic Interstellar Spaceflight."* Cambridge, MA: MIT Undergraduate Thesis. http://hdl.handle.net/1721.1/102711

Gilster, Paul. (2020). "The Interstellar Ramjet at 60." https://www.centauri-dreams .org/2020/04/03/the-interstellar-ramjet-at-60/

Heppenheimer, Thomas A. (1978). "On the Infeasibility of Interstellar Ramjets." *Journal of the British Interplanetary Society.* Vol. 31: 222.

Long, K. F., et al. (2010). "Project Icarus: Son of Daedalus – Flying Closer to Another Star." *Journal of the British Interplanetary Society.* Vol 62: 11.

O'Neill, Gerard K. (1977). *The High Frontier: Human Colonies in Space.* New York: William Morrow.

Sagan, Carl. (1963). "Direct Contact Among Galactic Civilizations by Relativistic Interstellar Spaceflight." *Planetary and Space Science.* Vol. 11: 485–498. https:// doi.org/10.1016/0032-0633(63)90072-2

Sagan, Carl. (1980). *Cosmos.* New York: Random House.

Schattschneider, Peter, and Jackson, Albert A. (2022). "The Fishback Ramjet Revisited." *Acta Astronautica.* Vol. 191: 227–234. https://doi.org/10.1016/j .actaastro.2021.10.039

Swinney, R. W., Freeland II, R. M., and Lamontagne, M. (2020). "Project Icarus: Designing a Fusion Powered Interstellar Probe." *Acta Futura.* Vol. 12: 47–59. https://doi.org/10.5281/zenodo.3747274

Whitmire, Daniel P. (1975). "Relativistic Spaceflight and the Catalytic Nuclear Ramjet." *Acta Astronautica.* Vol. 2: 497–509.

Whitmire, D., and Jackson, A. (1977). "Laser Powered Interstellar Ramjet." *Journal of the British Interplanetary Society.* Vol. 30: 223–226.

Winterberg, Friedwardt. (1972). "*Thermonuclear Micro-Bomb Rocket Propulsion.*" University of Nevada: Desert Research Institute. https://documents.theblack vault.com/documents/dtic/746027.pdf

Chapter 4: Project Orion

Air Force Contract AF 18(600)-1812. (1958). "Feasibility Study of a Nuclear Bomb Propelled Space Vehicle."

Brower, Kenneth. (1978). *The Starship and the Canoe*. New York: Harper Colophon.

Dyson, Freeman. (1968). "Interstellar Transport." *Physics Today*. (October). 41–45.

Dyson, Freeman. (1979). "The Greening of the Galaxy." In *Dyson, Freeman. Disturbing the Universe*. New York: Harper Colophon.

Dyson, Freeman. (1992). "Sir Phillip Roberts's Erolunar Collision." In Dyson, Freeman. *From Eros to Gaia*. New York: Pantheon.

Dyson, George. (2002). *Project Orion: The True Story of the Atomic Spaceship*. New York: Henry Holt.

Everett, C. J., and Ulam, S. M. (1955). "*On a Method of Propulsion of Projectiles by Means of External Nuclear Explosions.*" Los Alamos Scientific Laboratory of the University of California.

Heinlein, Robert. (1940). "Blowups Happen." *Astounding Science Fiction*. https://www.baen.com/Chapters/0743471598/0743471598___4.htm

Ley, Willy. (1968). *Rockets, Missiles, and Men in Space*. New York, Signet.

Taylor, T. B. (1957). *Note on the Possibility of Nuclear Propulsion of a Very Large Vehicle at Greater than Earth Escape Velocities*. General Atomic informal report GAMD-25 C.

Chapter 5: Where To?

Bailes, M., Lyne, A. G., and Shemar, S. L. (1991). "A Planet Orbiting the Neutron Star PSR 1829–10." *Nature*. Vol. 352: 311–313. https://doi.org/10.1038/352311a0

Deudney, Daniel. (2020). *Dark Skies: Space Expansionism, Planetary Geopolitics, and the Ends of Humanity*. New York: Oxford University Press.

Dick, Steven J. (1982). *Plurality of Worlds: The Extraterrestrial Life Debate from Democritus to Kant*. Cambridge, UK: Cambridge University Press.

Goldsmith, Donald, and Rees, Martin. (2022). *The End of Astronauts*. Cambridge, MA and London: The Belknap Press of Harvard University Press.

Goodyear, Dana. (2023). "Dangerous Designs." *The New Yorker*. (September 11).

Habitable Worlds Catalog. (2024). https://phl.upr.edu/projects/habitable-exo planets-catalog

Hernshaw, John. (2010). "Auguste Comte's Blunder." *Journal of Astronomical History and Heritage*. Vol. 13: 90–104.

Hershey, John L. (1973). "Astrometric Analysis of the Field of AC+65°6955 from Plates Taken With the Sproul 24-inch Refractor." *The Astronomical Journal*. Vol. 78: 421–425.

Kent, Bill. (2001). "Barnard's Wobble." *Swarthmore College Bulletin*. (March).

Lemonick, Michael D. (2012). *Mirror Earth: The Search for Our Planet's Twin*. New York: Bloomsbury.

Lyne, A. G., and Bailes, M. (1992). "No Planet Orbiting PSR 1829–10." *Nature*. Vol. 355: 213. https://doi.org/10.1038/355213b0

Mason, Christopher E. (2021). *The Next 500 Years: Engineering Life to Reach New Worlds*. Cambridge, MA: MIT Press.

Mayor, M., and Queloz, D. (1995). "A Jupiter-mass Companion to a Solar-type Star." *Nature*. 378: 355–359. https://doi.org/10.1038/378355a0

Van de Kamp, P. (1963). "Astrometric Study of Barnard's Star from Plates Taken with the 24-Inch Sproul Refractor." *The Astronomical Journal*. Vol. 68: 515–521.

Winn, Joshua N. (2019). "Who Really Discovered the First Exoplanet?" *Scientific American*. https://www.scientificamerican.com/blog/observations/who-really-discovered-the-first-exoplanet/

Wolszczan, A. and Frail, D. (1992). "A Planetary System Around the Millisecond Pulsar PSR1257+12." *Nature*. Vol. 355: 145–147.

Chapter 6: The World Ship

Bernal. J. D. (1929). *The World, the Flesh, and the Devil*. Foyle Publishing. https://www.quarkweb.com/foyle/WorldFleshDevil.pdf

Bond, A. and Martin, A. R. (1984). "World Ships – An Assessment of the Engineering Feasibility." *Journal of the British Interplanetary Society*. Vol. 37: 254.

Elhefnawy, Nader. (2008). "Revisiting Island One." *The Space Review*. (October 27).

Hein, A. M., Pak, M., Putz, D., Buhler, C., and Reiss, P. (2012) "World Ships: Architecture and Feasibility Revisited." *Journal of the British Interplanetary Society*. Vol. 65: 119–133.

Hein, A. M., Smith, C., Marin, F., and Staats, K. (2020). "World Ships: Feasibility and Rationale." *Acta Futura*. Vol. 12: 75–104.

Jain, V., et al. (2023). "Human Development and Reproduction in Space – A European Perspective." *NPJ Microgravity*. Vol. 9: 24. http://10.1038/s41526-023-00272-5

Marin, F., and Beluffi, C. (2018). "Numerical Constraints on the Size of Generation Ships." *Journal of the British Interplanetary Society*. Vol. 71: 382–393.

Martin, A. R. (1984). "World Ships – Concept, Cause, Cost, Construction and Colonisation." *Journal of the British Interplanetary Society*. Vol. 37: 243.

Matloff, Gregory L. (1976). "Utilization of O'Neill's Model 1 Lagrange Point Colony as an Interstellar Ark." *Journal of the British Interplanetary Society*. Vol. 29: 775–785.

McKendree, Thomas L. (1996). "Implications of Molecular Nanotechnology Technical Performance Parameters on Previously Defined Space System Architectures." *Nanotechnology.* Vol. 7: 204. https://10.1088/0957-4484/12/1/701

Miklavčič, Peter M., et al. (2022). "Habitat Bennu: Design Concepts for Spinning Habitats Constructed From Rubble Pile Near-Earth Asteroids." *Frontiers in Astronomy and Space Sciences.* Vol. 8. https://doi.org/10.3389/fspas.2021.645363

Moore, John H. (2003) "Kin-Based Crews for Interstellar Multi-Generation Space Travel." In Kondo, Yoji, et al. (eds.). *Interstellar Travel and Multi-Generation Space Ships.* Apogee Books.

O'Neill, Gerard K. (1974). "The Colonization of Space." *Physics Today.* (September): 32–39.

O'Neill, Gerard K. (1977). *The High Frontier: Human Colonies in Space.* New York: William Morrow.

Robinson, Kim Stanley. (2015). *Aurora.* New York: Orbit Books.

Salotti, J-M. (2020). "Minimum Number of Settlers for Survival on Another Planet." *Scientific Reports.* Vol. 16. http://doi.org/10.1038/s41598-020-66740-0

Smith, C. M. (2014). "Estimation of a Genetically Viable Population for Multigenerational Interstellar Voyaging." *Acta Astronautica.* Vol. 97: 16–29.

Chapter 7:   Hail Mary Propulsion Systems, Inc.

Anderson, Carl. (1932). "The Apparent Existence of Easily Deflectable Positives." *Science.* Vol. 76: 238–239.

Anderson, Carl. (1933). "The Positive Electron." *Physical Review.* Vol. 43: 491–494.

Anderson, Margot. (2016). "Antimatter Starship Scheme Coming to Kickstarter." *IEEE Spectrum.* (April 12).

Baur, G., et al. (1996). "Production of Antihydrogen." *Physics Letters B.* Vol. 368: 251–258.

CERN. (2011). "Making Antimatter." https://angelsanddemons.web.cern.ch/anti matter/making-antimatter.html#:~:text=Antimatter%20is%20produced% 20in%20many,antiprotons%20that%20can%20be%20trapped.

CERN. (n.d.). "The Story of Antimatter." https://timeline.web.cern.ch/taxonomy/ term/86?page=1

Clery, Daniel. (2022). "With Historic Explosion, a Fusion Breakthrough." *Science.* (December 13). https://www.science.org/content/article/historic-explosion-long-sought-fusion-breakthrough

Dirac, P. A. M. (1928). "The Quantum Theory of the Electron." *Proceedings of the Royal Society A.* Vol. 117: 610–624.

Dirac, P. A. M. (1931). "Quantised Singularities in the Electromagnetic Field. *Proceedings of the Royal Society A.* Vol. 133: 60. https://doi.org/10.1098/ rspa.1931.0130

Dorminey, Bruce. (2016). "Antimatter Space Propulsion Possible Within a Decade, Say Physicists." *Forbes.* (February 24).

Fecht, Sarah. (2016). "Former Fermilab Physicist Aims to Build a 'Star Trek'-Style Antimatter Engine." *Popular Science*. (March).

Forward, Robert L. (1984). "Roundtrip Interstellar Travel Using Laser-Pushed Lightsails." *Journal of Spacecraft and Rockets*. Vol. 21: 187–195.

Forward, Robert L. (1988). *Future Magic*. New York: Avon Books.

David, Leonard. (2022). "Nuclear Fusion Breakthrough: What Does It Mean for Space Exploration?" Space.com. December 15. https://www.space.com/nuclear-fusion-breakthrough-spacetravel

Heffernan, Virginia. (2023). "It's Time to Fall in Love With Nuclear Fusion – Again." *Wired*. (March 1).

ITER. (2023). "What Is ITER?" https://www.iter.org/proj/inafewlines

Jassby, Daniel. (2018). "ITER Is a Showcase … for the Drawbacks of Fusion Energy." *Bulletin of the Atomic Scientists*. (February 14).

Jassby, Daniel L. (2021). "Fusion Frenzy – A Recurring Pandemic." *Physics and Society*. Vol. 50 (October).

Jassby, Daniel. (2022). "The Quest for Fusion Energy." *Inference*. Vol. 7 (May).

Jassby, Daniel. (2022). "On the Laser Fusion Milestone." *Inference*. Vo. 7 (December).

Khatchadourian, Raffi. (2014). "A Star in a Bottle." *The New Yorker*. (March 3).

Kramer, David. (2022). "National Ignition Facility Surpasses Long-awaited Fusion Milestone." *Physics Today*. (December 13). https://doi.org/10.1063/PT.6.2.20221213a

Kramer, David. (2023). "National Ignition Facility Earns Its Name for a Second Time." *Physics Today*. (August 11). https://doi.org/10.1063/PT.6.2.20230811a

Mallove, Eugene F., and Matloff, Gregory L. (1989). *The Starflight Handbook*. New York: John Wiley & Sons.

Reich, Eugene Samuel. (2010). "Antimatter Held for Questioning." *Nature*. Vol. 468: 355. (November 18).

Schmidt, G. R., et al. (1999). "Antimatter Production for Near-Term Propulsion Applications." *35th Joint Propulsion Conference and Exhibit*. American Institute of Aeronautics and Astronautics.

Sharma, Shrijan. (2022). "Antimatter Propulsion and Its Application for Interstellar Travel: A Review." *International Journal for Research in Applied Science & Engineering Technology*. Vol. 10. (November). https://doi.org/10.22214/ijraset.2022.47427

Chapter 8:   The Fate of the Crew

Ahrari, Khulood, Omolaoye, T., Goswami, N., et al. (2022). "Effects of Space Flight on Sperm Function and Integrity." *Frontiers of Physiology*. (August 11). https://doi.org/10.3389/fphys.2022.904375

Arone, Alessandro, Ivaldi, T., Loganovsky, K., et al. (2021). "The Burden of Space Exploration on the Mental Health of Astronauts: A Narrative Review." *Clinical Neuropsychiatry*. Vol. 18: 237–246.

Carpenter, Alvin L. (2012). "The Non-Promise of Earthbound Religions into Space." *100YSS 2012 Public Symposium*, Houston, TX, September 14, 2012.

Cresswell, Matthew. (2012). "How Buzz Aldrin's Communion on the Moon Was Hushed Up." *The Guardian*. (September 13).

Garrett-Bakelman, Francine E., et al. (2019). "The NASA Twins Study: A Multidimensional Analysis of a Year-Long Human Spaceflight." *Science*. Vol. 364. http://doi:10.1126/science.aau8650

Kelly, Scott. (2017). *Endurance: A Year in Space, a Lifetime of Discovery*. New York: Knopf.

Krittanawong, Chayakrit, et al. (2023). "Human Health During Space Travel: State-of-the-Art Review." *Cell*. Vol. 12.: 40. https://doi.org/10.3390/cells12010040

Nelson, Mark. (1999). "Bioregenerative Recycling of Wastewater in Biosphere 2 Using a Constructed Wetland: 2-year Results." *Ecological Engineering*. Vol. 13: 189–197.

Nelson, Mark. (2017). "Setting the Record Straight About the Biosphere 2." Vice.com. (June 24).

Poynter, Jane. (2006). *The Human Experiment: Two Years and Twenty Minutes Inside Biosphere 2*. New York: Thunder's Mouth Press.

Scoles, Sarah. (2023). "Why We'll Never Live in Space." *Scientific American*. (October 1).

Weinersmith, Kelly, and Weinersmith, Zach. (2023). *A City on Mars*. New York: Penguin Press.

Zimmer, Carl. (2019). "The Lost History of One of the World's Strangest Science Experiments." *New York Times*. (March 29).

Chapter 9:   The Moral Status of the Trip

Cockell, Charles S. (2008). "An Essay on Extraterrestrial Liberty." *Journal of the British Interplanetary Society*. Vol. 61: 255–275.

Dyson, Freeman. (1968). "Interstellar Transport." *Physics Today*. (October). 41–45.

Hein, A. M., et al. (2012). "World Ships – Architectures and Feasibility Revisited." *Journal of the British Interplanetary Society*. Vol. 65: 119-133.

Hein, A. M., Smith, C., Marin, F., and Staats, K. (2020). "World Ships: Feasibility and Rationale." *Acta Futura*. Vol. 12: 75–104.

Flyvbjerg, Bent. (2014). "What You Should Know About Megaprojects and Why." *Project Management Journal*. Vol. 45: 6–19.

Martin, A. R. (1984). "World Ships – Concept, Cause, Cost, Construction and Colonisation." *Journal of the British Interplanetary Society*. Vol. 37: 243.

Regis, Edward. (1985.) "The Moral Status of Multigenerational Interstellar Exploration." In Finney, Ben R., and Jones, Eric M. (eds.). *Interstellar Migration and the Human Experience*. Berkeley, CA: University of California Press.

Robinson, Kim Stanley. (2015). *Aurora*. New York: Orbit Books.

Schwartz, James. (2018). "Worldship Ethics 101: The Shipborn." In *Proceedings of the 2017 Tennessee Valley Interstellar Workshop*. Vol. 71.

Smith, Cameron, and Davies, Evan. (2012). *Emigrating Beyond Earth: Human Adaptation and Space Colonization*. New York: Springer Praxis Books.

Thompson, Sarah G. (2003). "Language Change and Cultural Continuity on Multi-Generational Spaceships." In Kondo, Yoji, et al. (eds.). *Interstellar Travel and Multi-Generation Space Ships*. Apogee Books.

US Government Accountability Office. (2023). "Space Launch System: Cost Transparency Needed to Monitor Program Affordability." GAO-23-105609.

Chapter 10: Let Us Hibernate

Bradford, J., Talk, Doug, and Schaffer, Mark. (2014). "Torpor Inducing Transfer Habitat for Human Stasis to Mars." NASA Innovative Advanced Concepts (NIAC). Phase I. Final Report.

Bradford, John. (2023). Email to Ed Regis. December 29.

Choukér, Alexander, Ngo-Anh, Thu J., Biesbroek, Robin, et al. (2021). "European Space Agency's Hibernation (Torpor) Strategy for Deep Space Missions: Linking Biology to Engineering." *Neuroscience and Biobehavioral Reviews*. Vol. 131: 618–626. https://doi.org/10.1016/j.neubiorev.2021.09.054

Dyson, Freeman. (1968). "Human Consequences of the Exploration of Space." In Dyson, Freeman. (1992). *From Eros to Gaia*. New York: Pantheon Books.

Garber, Megan. (2012). "The Trash We've Left on the Moon." *Atlantic*. (December 19).

Goldsmith, Donald, and Rees, Martin. (2022). *The End of Astronauts*. Cambridge, MA and London: The Belknap Press of Harvard University Press.

Hayflick, L., and Moorhead, P. S. (1961). "The Serial Cultivation of Human Diploid Cell Strains." *Experimental Cell Research*. Vol. 25: 585–621. https://doi.org/10.1016/0014-4827(61)90192-6

Nespolo, R. F., Mejias C., and Bozinovic F. (2022). "Why Bears Hibernate? Redefining the Scaling Energetics of Hibernation." *Proceedings of the Royal Society B*. Vol. 289: 20220456. https://doi.org/10.1098/rspb.2022.0456

Chapter 11: Why Go?

Aldrin, Buzz, and Wachhorst, Wyn. (2004). "The Urge to Explore." *Mechanical Engineering*. Vol. 126. https://doi.org/10.1115/1.2004-NOV-2

Beason, Doug. (2002). "Interstellar Travel: Why We Must Go." In Kondo, Yoji, et al. (eds.). *Interstellar Travel and Multi-Generation Space Ships*. Apogee Books.

Collins, Michael. (1974). *Carrying the Fire*. New York: Farrar, Straus and Giroux.

Cronon, William. (1987). "Revisiting the Vanishing Frontier." *Western Historical Quarterly*. Vol. 18: 157–176.

Dyson, Freeman. (1992). *From Eros to Gaia*. New York: Pantheon Books.

Forward, Robert L. (1988). *Future Magic*. New York: Avon Books.

Freudenthal, Hans. (1960). *LINCOS: Design of a Language for Cosmic Intercourse*. Amsterdam: North-Holland Publishing Company.

Hickey, Raymond. (2023). *Life and Language Beyond Earth*. Cambridge, UK: Cambridge University Press. https://doi.org/10.1017/9781009229272

Kondo, Yoji. (2002). "Interstellar Travel and Multi-Generation Space Ships: An Overview." In Kondo, Yoji, et al. (eds.). *Interstellar Travel and Multi-Generation Space Ships*. Apogee Books.

Launius, Roger D. (2013). "Escaping Earth: Human Spaceflight as Religion." *Astropolitics: The International Journal of Space Politics and Policy*. Vol. 11: 45–64. http://10.1080/14777622.2013.801720

Limerick, P. (1992). "Imagined Frontiers: Westward Expansion and the Future of the Space Program." In Byerly, R. (ed.). *Space Policy Alternatives*, pp. 249–261. Boulder, CO: Westview Press.

Lin, Patrick. (2006). "Look Before Taking Another Leap for Mankind." *Astropolitics: The International Journal of Space Politics and Policy*. Vol. 4. http://10.1080/14777620601039701

Mason, Christopher E. (2021). *The Next 500 Years: Engineering Life to Reach New Worlds*. Cambridge, MA: MIT Press.

Poynter, Jane. (2006). *The Human Experiment: Two Years and Twenty Minutes Inside Biosphere 2*. New York: Thunder's Mouth Press.

Sagan, Carl. (1963). "Direct Contact Among Galactic Civilizations by Relativistic Interstellar Spaceflight." *Planetary and Space Science*. Vol. 11: 485–498. https://doi.org/10.1016/0032-0633(63)90072-2

Schwartz, James S. J. (2011). "Our Moral Obligation to Support Space Exploration." *Environmental Ethics*. 33: 67-88. https://doi.org/10.5840/enviroethics20113317

Turner, Frederick Jackson. (1893). "The Significance of the Frontier in American History." In Turner, Frederick Jackson. (1921). *The Frontier in American History*. New York: Henry Holt.

Weinersmith, Kelly, and Weinersmith, Zach. (2023). *A City on Mars*. New York: Penguin Press.

Wolfe, Tom. (1979). *The Right Stuff*. New York: Farrar, Straus and Giroux.

Chapter 12: The Odds

Abulafia, David. (2019). *The Boundless Sea*. New York: Oxford University Press.

Cooper, Ralph S. (1967). "Cristobal Columbo." In Clarke, Arthur C. (ed.). *The Coming of the Space Age*. New York: Meredith Press.

Deudney, Daniel. (2020). *Dark Skies: Space Expansionism, Planetary Geopolitics, and the Ends of Humanity*. New York: Oxford University Press.

Robinson, Kim Stanley. (2015). "*Our Generation Ships Will Sink*." In Bosch, T., and Scranton, R. (eds.). *What Future*. Unnamed Press.

INDEX